Service-Oriented Distributed Knowledge Discovery

Chapman & Hall/CRC
Data Mining and Knowledge Discovery Series

SERIES EDITOR

Vipin Kumar

University of Minnesota
Department of Computer Science and Engineering
Minneapolis, Minnesota, U.S.A.

AIMS AND SCOPE

This series aims to capture new developments and applications in data mining and knowledge discovery, while summarizing the computational tools and techniques useful in data analysis. This series encourages the integration of mathematical, statistical, and computational methods and techniques through the publication of a broad range of textbooks, reference works, and handbooks. The inclusion of concrete examples and applications is highly encouraged. The scope of the series includes, but is not limited to, titles in the areas of data mining and knowledge discovery methods and applications, modeling, algorithms, theory and foundations, data and knowledge visualization, data mining systems and tools, and privacy and security issues.

PUBLISHED TITLES

ADVANCES IN MACHINE LEARNING AND DATA MINING FOR ASTRONOMY
Michael J. Way, Jeffrey D. Scargle, Kamal M. Ali, and Ashok N. Srivastava

BIOLOGICAL DATA MINING
Jake Y. Chen and Stefano Lonardi

COMPUTATIONAL METHODS OF FEATURE SELECTION
Huan Liu and Hiroshi Motoda

CONSTRAINED CLUSTERING: ADVANCES IN ALGORITHMS, THEORY, AND APPLICATIONS
Sugato Basu, Ian Davidson, and Kiri L. Wagstaff

CONTRAST DATA MINING: CONCEPTS, ALGORITHMS, AND APPLICATIONS
Guozhu Dong and James Bailey

DATA CLUSTERING IN C++: AN OBJECT-ORIENTED APPROACH
Guojun Gan

DATA MINING FOR DESIGN AND MARKETING
Yukio Ohsawa and Katsutoshi Yada

DATA MINING WITH R: LEARNING WITH CASE STUDIES
Luís Torgo

FOUNDATIONS OF PREDICTIVE ANALYTICS
James Wu and Stephen Coggeshall

GEOGRAPHIC DATA MINING AND KNOWLEDGE DISCOVERY, SECOND EDITION
Harvey J. Miller and Jiawei Han

HANDBOOK OF EDUCATIONAL DATA MINING
Cristóbal Romero, Sebastian Ventura, Mykola Pechenizkiy, and Ryan S.J.d. Baker

INFORMATION DISCOVERY ON ELECTRONIC HEALTH RECORDS
Vagelis Hristidis

INTELLIGENT TECHNOLOGIES FOR WEB APPLICATIONS
Priti Srinivas Sajja and Rajendra Akerkar

INTRODUCTION TO PRIVACY-PRESERVING DATA PUBLISHING:
CONCEPTS AND TECHNIQUES
Benjamin C. M. Fung, Ke Wang, Ada Wai-Chee Fu, and Philip S. Yu

KNOWLEDGE DISCOVERY FOR COUNTERTERRORISM AND LAW ENFORCEMENT
David Skillicorn

KNOWLEDGE DISCOVERY FROM DATA STREAMS
João Gama

MACHINE LEARNING AND KNOWLEDGE DISCOVERY FOR
ENGINEERING SYSTEMS HEALTH MANAGEMENT
Ashok N. Srivastava and Jiawei Han

MINING SOFTWARE SPECIFICATIONS: METHODOLOGIES AND APPLICATIONS
David Lo, Siau-Cheng Khoo, Jiawei Han, and Chao Liu

MULTIMEDIA DATA MINING: A SYSTEMATIC INTRODUCTION TO CONCEPTS AND THEORY
Zhongfei Zhang and Ruofei Zhang

MUSIC DATA MINING
Tao Li, Mitsunori Ogihara, and George Tzanetakis

NEXT GENERATION OF DATA MINING
Hillol Kargupta, Jiawei Han, Philip S. Yu, Rajeev Motwani, and Vipin Kumar

RELATIONAL DATA CLUSTERING: MODELS, ALGORITHMS, AND APPLICATIONS
Bo Long, Zhongfei Zhang, and Philip S. Yu

SERVICE-ORIENTED DISTRIBUTED KNOWLEDGE DISCOVERY
Domenico Talia and Paolo Trunfio

SPECTRAL FEATURE SELECTION FOR DATA MINING
Zheng Alan Zhao and Huan Liu

STATISTICAL DATA MINING USING SAS APPLICATIONS, SECOND EDITION
George Fernandez

TEMPORAL DATA MINING
Theophano Mitsa

TEXT MINING: CLASSIFICATION, CLUSTERING, AND APPLICATIONS
Ashok N. Srivastava and Mehran Sahami

THE TOP TEN ALGORITHMS IN DATA MINING
Xindong Wu and Vipin Kumar

UNDERSTANDING COMPLEX DATASETS:
DATA MINING WITH MATRIX DECOMPOSITIONS
David Skillicorn

Service-Oriented Distributed Knowledge Discovery

Domenico Talia
Paolo Trunfio

CRC Press
Taylor & Francis Group
Boca Raton London New York

CRC Press is an imprint of the
Taylor & Francis Group, an **informa** business

A CHAPMAN & HALL BOOK

CRC Press
Taylor & Francis Group
6000 Broken Sound Parkway NW, Suite 300
Boca Raton, FL 33487-2742

© 2013 by Taylor & Francis Group, LLC
CRC Press is an imprint of Taylor & Francis Group, an Informa business

Printed in the United States of America on acid-free paper
Version Date: 20120822

International Standard Book Number: 978-1-4398-7531-5 (Hardback)

Library of Congress Cataloging-in-Publication Data

Talia, Domenico.
    Service-oriented distributed knowledge discovery / Domenico Talia, Paolo Trunfio.
        p. cm. -- (Chapman & Hall/CRC data mining and knowledge discovery series)
    Includes bibliographical references and index.
    ISBN 978-1-4398-7531-5 (hardback)
    1. Data mining. 2. Service-oriented architecture (Computer science) I. Trunfio, Paolo.
II. Title.

QA76.9.D343T3484 2012
006.3'12--dc23   2012028117

Visit the Taylor & Francis Web site at
http://www.taylorandfrancis.com

and the CRC Press Web site at
http://www.crcpress.com

*To Enza, Francesco, and Marianna for the
life we are writing together.*

DT

*To Thomas who was born during the writing of the book,
and Erika with whom I lived the great joy of his birth.*

PT

Contents

Preface

Data analysis techniques and services are needed to mine the massive amount of data available and to extract useful knowledge from it. The service-oriented architecture (SOA) is used today as a model to develop software systems as a collection of services that are units of functionality and are interoperable in an open programming scenario. Service-oriented architectures can offer tools, techniques, and environments to support analysis, inference, and discovery processes over large data repositories available in many scientific and business areas. Knowledge discovery services, based on the availability of huge operation and application data and on the exploitation of data mining techniques, support and enable large-scale knowledge discovery applications on service-oriented architectures such as Web servers, Grids, and Cloud computing platforms.

This new approach can be referred to as service-oriented knowledge discovery. It addresses issues related to distributed knowledge discovery algorithms, data services composition, data and knowledge integration, and service-oriented data mining workflows, which provide the main components for extracting useful knowledge from the often unmanageable data volumes available today from many sources. This is done by exploiting data mining and machine learning distributed models and techniques in service-oriented infrastructures.

This book is about distributed knowledge discovery techniques, algorithms, and systems based on the service-oriented paradigm. It introduces service-oriented knowledge discovery techniques, models, and architectures and explains how those can also be implemented through a detailed description of real software systems that are addressed in some chapters. The final part of the book illustrates distributed knowledge discovery applications and discusses the future role of service-oriented knowledge discovery in ubiquitous discovery processes and in large-scale data analytics.

The book covers several key areas in data mining and service-oriented computing: (1) concepts and principles of distributed knowledge discovery and service-oriented data mining; (2) design of services for data analytics; (3) real systems for implementing distributed knowledge discovery applications; (4) mobile data mining models; and (5) future trends in service-oriented data analytics.

The book is for researchers, graduate students, and practitioners in data mining, knowledge discovery, and service-oriented computing fields. Researchers will find some of the latest achievements in the area and many examples of the state-of-the-art in service-oriented knowledge discovery. Both readers who are beginners to the subject and experienced readers in the distributed data mining domain will find topics of interest. Furthermore, graduate students and young researchers will learn useful concepts related to distributed data mining and service-oriented data analysis.

Several sections contain real applications and case studies that provide useful information for developers about issues, opportunities, and successful approaches in the practical use of service-oriented knowledge discovery in databases (KDD) frameworks, for example, the Knowledge Grid and Weka4WS. The chapters are written in such a way that the book can also be used as a reference text in graduate and postgraduate courses in distributed knowledge discovery.

People from the publisher, Taylor & Francis/CRC Press, particularly Jennifer Ahringer, Randi Cohen, and Linda Leggio, must be commended for their support and work during the publication process.

We hope readers will find the book content interesting and useful, as we found it interesting and exciting to write.

Domenico Talia
University of Calabria and ICAR-CNR
Rende, Italy

Paolo Trunfio
University of Calabria
Rende, Italy

Acknowledgments

We would like to thank the colleagues who cooperated with us on the research themes of this book. In particular, we would like to thank Mario Cannataro, Eugenio Cesario, Carmela Comito, Antonio Congiusta, Marco Lackovic, and Oreste Verta.

Authors

Domenico Talia is a full professor of computer engineering at the University of Calabria, Italy, and the director of the Institute of High Performance Computing and Networking of the Italian National Research Council (ICAR-CNR). His research interests include parallel and distributed data mining algorithms, Cloud computing, Grid services, distributed knowledge discovery, peer-to-peer systems, and parallel programming models. Talia has published 6 books and about 300 papers in archival journals such as *Communications of the ACM, IEEE Computer, IEEE TKDE, IEEE TSE, IEEE TSMC-A, IEEE TSMC-B, IEEE Micro, ACM Computing Surveys, FGCS, Parallel Computing, IEEE Internet Computing,* and conference proceedings. He is a member of the editorial boards of *IEEE Transactions on Computers,* the *Future Generation Computer Systems* journal, the *International Journal of Web and Grid Services,* the *Journal of Cloud Computing—Advances, Systems and Applications,* the *Scalable Computing Practice and Experience* journal, the *International Journal of Next-Generation Computing, Multiagent and Grid Systems: An International Journal,* and the *Web Intelligence and Agent Systems* journal. He was guest editor of special issues of *IEEE Transactions on Software Engineering, Parallel Computing,* and *Future Generation Computer Systems,* and served as a program chair or program committee member of several conferences. Talia is a member of the Association for Computing Machinery (ACM) and IEEE Computer Society.

Paolo Trunfio is an assistant professor of computer engineering at the University of Calabria, Italy. He received his Ph.D. in systems and computer engineering from the same university. In 2007, he was a visiting researcher at the Swedish Institute of Computer Science (SICS) in Stockholm. Previously, he was a research collaborator at the Institute of Systems and Computer Science of the Italian National Research Council

(ISI-CNR). His research interests include Grid computing, Cloud computing, service-oriented architectures, distributed knowledge discovery, and peer-to-peer systems. Trunfio is a member of the editorial board of the *ISRN Artificial Intelligence* journal, and has served as program committee member for several conferences. He is author of about 80 papers published in books, conference proceedings, and international journals such as the *Journal of Computer and System Sciences*, the *Journal of Parallel and Distributed Computing*, *Concurrency and Computation: Practice and Experience*, *Future Generation Computer Systems*, *IEEE TSMC-B*, *IEEE Internet Computing*, and *Communications of the ACM*.

CHAPTER 1

Distributed Knowledge Discovery

An Overview

W E INTRODUCE HERE THE BASIC CONCEPTS of knowledge discovery and data mining. Models and techniques are presented together with the main approaches used in parallel and distributed knowledge discovery. In particular, in Section 1.1 we introduce the main motivations of knowledge discovery and data mining, discuss some examples, and describe the knowledge discovery process models. Section 1.2 discusses the most important data mining techniques, and Section 1.3 analyzes different forms of parallelism that can be exploited in data mining algorithms. Finally, Section 1.4 focuses on knowledge discovery and data mining in distributed environments.

1.1 KNOWLEDGE DISCOVERY AND DATA MINING CONCEPTS

The ability to generate, acquire, and store digital data is increasing rapidly due to technology evolution and increasing computerization of commercial, scientific, and administrative domains. It has been estimated that the amount of data stored in worldwide databases doubles every 20 months. This explosive growth has generated an urgent need for new technologies and automatic tools that can effectively aid in the transformation of this huge amount of data into useful information and knowledge. Such information may provide scientists, engineers, businesspersons, and citizens with a vast new resource that can be exploited, for example, to make scientific discoveries, optimize industrial systems, and uncover financially valuable patterns. Generally speaking, the information and knowledge that can be extracted from that very large amount of data can improve the quality of life of humans; therefore, it is vital to provide tools that may help in doing this.

There is a significant need for a new generation of techniques and tools capable of assisting users to intelligently and automatically analyze these "mountains" of data in order to find useful knowledge. These techniques and tools are the subject of study in the field of knowledge discovery in databases (KDD). A *knowledge discovery system* is a system that finds knowledge that it previously did not—that is, it was not implicit in its algorithms or explicit in its representation of domain knowledge. The term *knowledge discovery in databases* was coined to refer to the whole process of knowledge discovery in data and emphasize the application, at a high level, of particular data analysis methods known under the name of *data mining*. KDD is therefore considered the whole process, typically interactive and iterative, to find and interpret data relationships that involve repeated applications of specific data mining methods and algorithms, and the interpretation of the reports generated by these algorithms.

Data mining techniques play a key role in the KDD process as they automatically discover and extract information that is not readily visible from a large amount of data (Fayyad and Uthurusamy, 1996). In general, data mining tasks identify and characterize relations between data, without necessarily requiring the user to formulate specific questions. Data mining algorithms search for patterns, trends, associations, and correlations between data, and highlight the information considered potentially valuable for users. As data mining techniques extract previously unknown information and rules rather than deny or prove hypotheses, they allow the discovery of knowledge, for instance indentifying trends or signaling risks. Therefore, they differ from other technologies, such as data warehouse and online analytical processing (OLAP), that allow only the verification of the available knowledge by storing and extracting data for efficient and easier reporting. Data mining is used in many classes of applications, such as risk analysis in finance and banking, tax fraud detection, analysis of the behavior of customers to purchase goods (*market basket analysis*), planning of advertising campaigns targeted at particular classes of customers, and analysis of scientific data in astronomy, biology, and medicine. In particular, data mining techniques are used in the area of market basket analysis to discover associations between items purchased by a customer during a single visit to a shop. In this context, an example of association rule is as follows:

70% of customers who buy spaghetti also buys tomatoes.

Through the discovery of such rules, processes such as procurement planning (orders), allocation of products in the store, and marketing campaigns can be improved. These types of applications are widely used by supermarket chains and are, in many cases, allowed to identify associations between products not previously known.

Another significant application of data mining techniques is the granting of a loan to a customer by a bank. In this case, the goal is to characterize the customers to whom credit should be granted based on the likelihood that they will honor the debt to be contracted. Data mining algorithms allow analyzing the bank's customers' data to define risk classes of customers according to information such as income, credit/debit already contracted, the amount of withdrawals in a given period, and so on. Prior to granting new credit, the bank can use this classification to decide which class a requesting client belongs to, and then determine the risk of the credit requested.

Many other significant examples of data mining applications can be found in both commercial and scientific areas. For example, data mining algorithms are used for detection of quasars by distinguishing them from other stars and galaxies. The mining algorithms are able to identify distant quasars based on their brightness correlated to their distance, which is typically 10 billion light-years away. Data mining may help in identifying false positives in image screening, with which environmental scientists detect oil stains from the image analysis of the coastal waters provided by radar satellites. In power load forecasting, a crucial problem in the electric power industry, data mining allows accurate power demand estimates to be obtained more rapidly. In the preventive maintenance of electromechanical appliances such as engines and generators, data mining helps specialists in the diagnosis of the failure type in order to prevent the interruption of industrial processes.

A fundamental aspect of the KDD process, and of data mining in particular, is that in order to obtain a better generalization it is necessary to have a better understanding. Improvements in knowledge may be obtained by collecting more information, defining more complex models, or through a more in-depth theoretical analysis (Chen, Han, and Yu, 1996).

As mentioned above, with the large quantity of data available in digital format, high-performance algorithms are necessary. For this reason, the use of advanced techniques is of great importance to create KDD and data mining systems. The analysis capabilities of data mining algorithms, coupled with *data warehousing* and *database management* technology, allow companies, banks, research centers, and governments to gain information and knowledge to better support their information and decision processes, providing them with significant advantages over competitors who do not benefit from this technology.

1.1.1 Knowledge Discovery in Databases (KDD) Process Models

A KDD process model consists of a set of processing steps that can be followed by practitioners when they execute their data analysis projects in order to help plan, work through, and reduce costs by detailing procedures to be performed in each of the steps. As mentioned earlier, data mining is described to be a part of the KDD process, as initially proposed by Fayyad, Piatetsky-Shapiro, and Smyth (1996). The approach to gain knowledge out of a set of data was separated by Fayyad and colleagues into five steps (see Figure 1.1):

1. *Selection*: Relevant information is selected from structured or unstructured data sources such as databases, file systems, or Web servers. A target dataset is created, focusing on a subset of variables or data samples, on which discovery is to be performed.

2. *Preprocessing*: Unimportant or erroneous elements of the provided data are detected and filtered out. The less noise contained in data, the higher will be the accuracy of the results of data mining. Elements of preprocessing span from the cleaning of wrong data

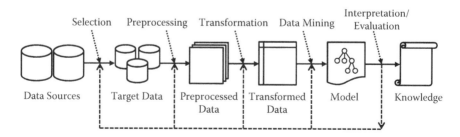

FIGURE 1.1 A representation of the knowledge discovery in databases (KDD) process.

(data inconsistencies such as out-of-range values like "Age = –10" or impossible data combinations like "Gender = male, Pregnant = yes"), the treatment of missing values, and the creation of new attributes.

3. *Transformation*: Data are transformed into a format compatible to data mining algorithms using dimensionality reduction or transformation methods.

4. *Data mining*: This stage consists of the searching for patterns of interest in a particular representational form, depending on the data mining objective (usually, prediction or classification).

5. *Interpretation/evaluation*: This stage reveals whether or not the discovered pattern is interesting—that is, whether it contains knowledge at all. Because of this, it is important to represent the result in an appropriate way so it can be examined thoroughly. If the located pattern is not interesting, then it is important to study the cause for it and possibly fall back on a previous step to make an attempt in another direction.

Other modified approaches to analyze the data followed the original proposal. For example, the former called 5A's (Assess, Access, Analyze, Act, and Automate) has been proposed by SPSS Inc., or SEMMA (Sample, Explore, Modify, Model, Assess), proposed by SAS Institute Inc. to be used in the Enterprise Miner. Both of these solutions though are closely connected to vendor products.

A standard that is industry-neutral, tool-neutral, and application-neutral is CRISP-DM (Cross-Industry Standard Process for Data Mining) (Chapman et al., 2000), developed by analysts representing DaimlerChrysler, SPSS, and NCR to provide a nonproprietary and freely available standard process for fitting data mining into the general problem-solving strategy of a business or research unit. A poll of data mining practitioners, conducted by KDnuggets in 2002,[*] 2004,[†] and 2007,[‡] revealed that CRISP-DM is the leading methodology used by data miners. It consists of a cycle with six stages:

---

[*] KDnuggets, 2002, July, Polls: What Main Methodology Are You Using for Data Mining? http://www.kdnuggets.com/polls/2002/methodology.htm (accessed June 2012).

[†] KDnuggets, 2004, April, Polls: Data Mining Methodology, http://www.kdnuggets.com/polls/2004/data_mining_methodology.htm (accessed June 2012).

[‡] KDnuggets, 2007, August, Polls: Data Mining Methodology, http://www.kdnuggets.com/polls/2007/data_mining_methodology.htm (accessed June 2012).

1. *Business understanding*: The project objectives and requirements are enunciated clearly in terms of the business or research unit as a whole. These goals and restrictions are translated into the formulation of a data mining problem definition, and a preliminary strategy for achieving these objectives is prepared.

2. *Data understanding*: Initial data are collected and exploratory data analysis is used to familiarize with the data and discover initial insights, the quality of the data is evaluated, and interesting subsets are selected that may contain actionable patterns.

3. *Data preparation*: The final dataset to be used for all subsequent phases is prepared from the initial raw data, the appropriate cases and variables to be analyzed are selected for the analysis, transformations are performed on certain variables, and if needed, the raw data are cleaned so that it is ready for the modeling tools.

4. *Modeling*: Appropriate modeling techniques are selected and applied, and model settings are calibrated to optimize results; if necessary, users loop back to the data preparation phase to bring the form of the data in line with the specific requirements of a particular data mining technique.

5. *Evaluation*: one or more models delivered from the modeling phase are evaluated for quality and effectiveness before deploying them for use in the field. It is determined whether the model in fact achieves the objectives set for it in the first phase and established whether some important facet of the business or research problem has not been accounted for sufficiently.

6. *Deployment*: The models created are used.

1.2 DATA MINING TECHNIQUES

As mentioned before, data mining is the core step of the KDD process that aims at discovering useful patterns or models in data. In fact, the goal of data mining is to find a (possibly complex) mathematical or logical description of relationships and patterns in a dataset. The data mining process attempts, therefore, to understand the similarities of the available data using a simplification of them, called a *model*. This model plays the role of inferred knowledge. Several data mining algorithms have been defined in literature in different research fields such as statistics, machine learning, mathematics, artificial intelligence, pattern recognition, and databases, each of which uses specialized techniques for the respective application field.

A database can be regarded as a reliable store of information. One of the main purposes of this store is to retrieve information efficiently. This information is not just the set of stored data, but more generally it is the information that can be inferred from the database. From a logical point of view, there are basically two main inference techniques: *deduction* and *induction*.*

The deduction technique allows information to be inferred that is a logical consequence of the information contained in the database. Many database management systems (DBMSs), such as relational systems, provide simple operators for information deduction, such as the *join* operator. In this field, however, there was a significant development for the construction of deductive databases that enrich a DBMS with logical deduction capabilities (*knowledge base management systems*, KBMSs).

The induction allows information to be inferred that is a generalization of the information contained in the database. This type of information constitutes knowledge (i.e., general statements about properties of objects). Hence, the inductive process looks for regularities in the database (i.e., combinations of values of certain attributes that are shared by all the facts in the database). These regularities can be formulated in terms of rules, so that the value of an attribute can be predicted in terms of other attributes.

The difference between deduction and induction is that while deduction infers data that can always be found to be valid within the database, provided of course that the database is correct, induction infers assertions that are supported only by the current database but may not be valid in general. Therefore, a primary task of induction is to select the rules most likely to minimize errors in the case they have to be tried on facts previously unknown.

As mentioned earlier, many DBMSs support deductive inference (KBMS). In recent years, there has been a growing interest in extending DBMSs to support inductive inference (i.e., with capabilities of extracting useful knowledge implicitly contained in the database, which is exactly the main objective of data mining) (Holsheimer and Siebes, 1994).

---

* There is also another type of inference, *abduction* (Charniak and McDermott, 1985), which infers hypotheses that explain observed events. For a discussion on the relationship between induction and abduction, see Dimopoulos and Kakas (1996).

Inductive learning, whose automation is the subject of research in the *machine learning* area, can be considered as an attempt by intelligent beings (often called *cognitive systems*) to understand their environment through a simplified view of that environment, consisting in the definition of a model. During the learning phase, the cognitive system observes its environment and recognizes in it similarities between objects and events. Then, the system groups similar objects into classes and builds rules to predict the behavior of the classes' components. The data mining process can therefore be seen as a process of inductive learning when the environment is a database.

1.2.1 Components of Data Mining Algorithms

In general, data mining algorithms can be viewed as the composition of certain building blocks and principles. In particular, data mining algorithms are made from a combination of three components: the *model*, the *model evaluation*, and the *search method*.

Model: In general, a model is an abstraction of the reality of interest. For the definition of a model, two factors must be taken into account in data mining: the *task* of the model and its *representation*. In general, the task can be of two types: *prescription* and *description*. The *prescription* involves using some variables or fields to predict unknown or future values of other variables of interest, and is one of the main goals of machine learning and pattern recognition applications. The *description* focuses on finding relationships interpretable by the users that describe the data. In any case, the most common types of tasks carried out in the data mining algorithms include the following:

- *Classification*: Classifies a dataset in one or more predefined classes so that the classification is able to predict the membership of a data instance to a certain class from a given set of predefined classes. For instance, a set of outlet store clients can be grouped in three classes: high-spending, average-spending, and low-spending clients.

- *Regression*: Associates a dataset to a variable and predicts the value of that variable. There are many applications of regression, such as assessing the likelihood that a patient can get sick from the results of diagnostic tests and predicting the margin of victory of a sports team based on the results and technical data of the previous matches.

- *Clustering*: Tries to identify a finite set of categories or groupings (*clusters*) to describe the data. The categories can be mutually exclusive and exhaustive or can consist of a more extensive representation, such as in the case of hierarchical categories. Examples of clustering applications in KDD concern the discovery of homogeneous subsets of consumers in a database of commercial sales. Unlike classification in which classes are predefined, in clustering the classes must be derived from data, looking for clusters based on metrics of similarity between data without user support.

- *Summarization*: Provides a compact description of a subset of data. One example is the tabulation of the mean and standard deviation of each field. More complex functions involve summary rules, the discovery of functional relationships between variables. Summarization techniques are often used in the interactive analysis of data and the automatic generation of reports.

- *Dependency modeling*: Consists of finding a model that describes significant dependencies between variables. Dependency models are at two levels: the structural level that specifies, often in graphic form, which variables are locally dependent from each other, and the quantitative level that specifies the power of dependencies using a numerical scale.

- *Relationship analysis* or *association*: Determines relationships between fields in a database in order to derive multiple correlations that meet the specified thresholds. For example, association rules describe which products are commonly sold with other products in one or more stores.

- *Episode discovery* or *prediction:* Looks for relationships between temporal sequences of events. For example, if a given event e_1 happens, then within a time interval *t,* there will be an event e_2 with probability *p.*

Model representation is the language used to describe the patterns to be discovered. Model representation determines both the flexibility of the model in representing data and the interpretability of the model in human terms. Typically, more complex models may better represent the data but may be more difficult to understand. The most popular model representations are decision trees, production rules, neural networks, genetic algorithms, methods based on examples, probabilistic dependency models, and relational models.

Model evaluation: Determines how much a particular model and its parameters satisfy the criteria of the KDD process. The model evaluation then establishes the criterion for preference of one model over another on the basis of statistics or logic criteria.

Search method: A search method consists of two components: *parameter search* and *model search*. In the first case, the algorithm must search for the parameters that optimize the evaluation criteria of the model, based on the observed dataset and a representation of the model. The model search can therefore be considered as an iterative process over the parameter search method. A family of models is considered, and each one is evaluated by the parameter search method. Because the space of possible models is generally very large, the implementations of search methods use heuristic search techniques.

1.2.2 Model Representation

Model representation is the language used to describe discoverable patterns. If the representation language is too limited, the accuracy of the model will also be low. In the following, the main representation languages are described.

1.2.2.1 Decision Trees

Decision tree learning is one of the most developed techniques for partitioning sets of objects into classes. Decision trees were first introduced in Quinlan's ID3 learning system (Quinlan, 1986), one of the earliest learning systems. A decision tree classifies examples into a finite number of predefined classes. The tree nodes are labeled with the names of attributes, the arcs are labeled with the possible values of the attribute, and the leaves are labeled with the different classes. An object is classified by following a path along the tree formed by the arcs corresponding to the values of its attributes.

Suppose we have a training set consisting of tuples that describe the weather conditions at a particular time to determine whether or not to play beach volleyball. Tuples contain information about the weather, which can be *sunny*, *cloudy*, or *rainy*; or the humidity, which can be *high*

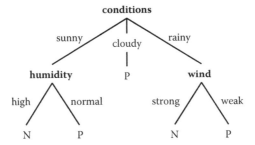

FIGURE 1.2   Example of a decision tree.

or *normal*; and the wind, which can be *strong* or *weak*. In addition, each tuple contains an additional attribute that classifies it as belonging to the class *P* (*play*) or *N* (*not play*), on the basis of the values of the other attributes. The goal of classification is to build a simple tree that classifies all tuples in the training set correctly. We could, for example, build the tree shown in Figure 1.2.

1.2.2.2 Production Rules

One of the main disadvantages of decision trees is that they tend to grow too much in real-world applications and are therefore difficult to interpret. A substantial research effort has been made to transform decision trees into simpler representations.

Quinlan (1987) described a technique for generating production rules from decision trees. This transformation is useful for three different reasons. First, production rules are a formalism widely used to represent knowledge, and second, they are easily interpreted by the experts because of their high modularity; each rule can be understood without reference to other rules. Finally, the accuracy of classification of a decision tree can be improved by transforming the tree into a set of production rules that contain a smaller number of attribute-value conditions. In fact, some conditions may be redundant, and it has been shown that their removal reduces the rate of misclassification on the examples that are not part of the training set (Quinlan, 1987).

For example, the decision tree of Figure 1.2 can be transformed into the following set of rules:

if conditions = sunny *and* humidity = high *then* class = N
if conditions = rainy *and* wind = strong *then* class = N
default class = P

1.2.2.3 Decision Lists

Decision lists, proposed in Rivest (1987), are a generalization of decision trees. A decision list is a list of pairs:

$$(\phi_1, C_1), \ldots, (\phi_r, C_r)$$

where each ϕ_i is a basic description, each C_i is a class, and the last description ϕ_r is the constant *true*. The class of an object o is C_j if j is the smallest index of a description ϕ_j that covers object o. This index always exists because the last term is always true—that is the default class. A decision list may be thought of as a formalism to define classes by providing the general patterns with some exceptions. The exceptions are the first terms of the list, while the latter terms are more general descriptions. For example, the following decision list could be used to describe the classes of sedans and utility cars:

(carmaker = Skoda Λ price = low, sedan)
(price = low, utility car)
(true, sedan)

1.2.2.4 Association Rules

Agrawal, Imielinski, and Swami (1993) introduced association rule learning or discovery as a method for discovering interesting relations between variables in large datasets. The problem of association rule mining is defined as follows: Let $I = \{i_1, i_2, \ldots, i_n\}$ be a set of n binary attributes called *items*. Let $D = \{t_1, t_2, \ldots, t_m\}$ be a set of transactions called the *dataset*. Each transaction in D has a unique transaction ID and contains a subset of the items in I. A *rule* is defined as an implication of the form $X \Rightarrow Y$, where $X, Y \subseteq I$, and $X \cap Y = \emptyset$. The sets of items (*itemsets*, for short) X and Y are called *antecedent* and *consequent* of the rule, respectively.

To illustrate the concepts, we use a simple example from the supermarket domain. An example rule for the supermarket could be (*pasta, tomatoes*) $\Rightarrow$ (*olive oil*), meaning that if customers buy pasta and tomatoes, they also buy olive oil.

In addition to the antecedent and the consequent, an association rule has two numbers that express the degree of uncertainty about the rule. In association rule learning, the antecedent and consequent are sets of items that are disjoint. The first number is called the *support* for the rule. The support is the number of transactions that include all items in the antecedent and consequent parts of the rule. The support can be expressed as a percentage of the total number of records in the dataset. For example, if an itemset {*pasta, tomatoes, olive oil*} occurs in 20% of all transactions (one out of five transactions), it has a support of 1/5 = 0.2.

The second number is known as the *confidence* of the rule. Confidence is the ratio of the number of transactions that include all items consequent as well as antecedent (the support) to the number of transactions that include all items in the antecedent: *confidence*($X \Rightarrow Y$) = *support*($X \cup Y$)/ *support*(X). For example, if a supermarket database has 200,000 transactions, out of which 4,000 include both items X and Y and 1,000 of these include item Z, the association rule "If X and Y are purchased, then Z is purchased at the same time" has a support of 1,000 transactions (alternatively 0.2% = 1,000/200,000) and a confidence of 25% (= 1,000/4,000).

Association rules are usually required to satisfy a user-specified *minimum support* and a user-specified *minimum confidence* at the same time. Association rule generation is usually split up into two separate steps:

First, minimum support is applied to find all *frequent itemsets* in a dataset.

Second, these *frequent itemsets* and the minimum confidence constraint are used to form rules.

While the second step is straightforward, the first step needs more attention. Finding all frequent itemsets in a dataset is difficult because it involves searching all possible itemsets (item combinations). The set of possible itemsets is the power set over I and has size 2^{n-1} (excluding the empty set that is not a valid itemset). Although the size of the power set grows exponentially in the number of items n in I, an efficient search is possible using the *downward-closure property* of support (also called *anti-monotonicity*), which guarantees that for a frequent itemset, all its subsets are also frequent; therefore for an infrequent itemset, all its supersets must also be infrequent. Exploiting this property, efficient algorithms such as Apriori (Agrawal and Srikant, 1994) and Eclat (Zaki, 2000) can find all frequent itemsets.

1.2.2.5 Neural Networks

Artificial neural networks (usually called *neural networks*) have a structure and a computational model very different from the processing models defined in artificial intelligence. The idea behind the research on neural networks is that intelligent systems can be obtained using a computational scheme somewhat similar to that of the human brain. To simulate the human brain model, neural networks are composed of a large number of *artificial neurons* that communicate with each other through a set of connections that play the role of *synapses*. It should be noted, however, that neural networks represent only a very simplified model of the human brain.

In neural networks, unlike almost all computational models, there is not a program that specifies the operations to be executed, but the computation is defined through the characteristics of processing units (artificial neurons) and their interconnections (Fogelman Soulie, 1991; McClelland and Rumelhart, 1986). A neural network learns through experience rather than through a program that specifies its behavior. This is the fundamental difference over the other approach of artificial intelligence based on the manipulation of symbols through the use of rules that form the program to execute.

The main features of a neural network are:

- A high number of simple processing units (artificial neurons)

- A high number of connections between neurons (synapses)

- A parallel and distributed control scheme

- A learning algorithm

The goal of a neural network is to associate an output y to a set of input signals $x_1, x_2, ..., x_n$. In a neural network a global controller of the computation state does not exist. Each artificial neuron computes its state y on the basis of local information (i.e., evolves according to its own law determined by a transition function f that specifies the dependence of state y from the signals received through the connections). The set of states of all the neurons constitutes the global network state. In general, f is a nonlinear function of the linear combination of the inputs through some coefficients that are given by the weights (p_{ij}) associated with the neuron connections:

$$y_i = f(\Sigma_j p_{ij} x_j)$$

As mentioned earlier, one of the most interesting features of neural networks is the ability to learn from experience, unlike classical artificial intelligence systems where knowledge of the system must be explicitly programmed (Fahlman and Hinton, 1987). Many learning algorithms have the characteristic of strengthening the connections between elements that have the same values of activation in the coherent state to be stored. In particular, the learning rules specify an initial set of weights associated to the connections and indicate how these weights should be adjusted to improve network performance (Simpson, 1990; Wasserman, 1989). Learning can take place, at least in part, by providing to the neural network a number of examples that constitute the training set. Only after the network has learned to respond correctly to all the inputs in the training set, is it ready to be used to solve the problems for which it was designed.

The neural network models proposed so far can be divided into four categories (Willshaw, 1988):

- *Auto-associators* combine the stored patterns themselves. These neural networks show in output a pattern *P*, which is stored in them, whenever it is provided as input a pattern similar enough to *P*. For example, if a pattern with "noise" or partial is provided, the network returns the complete pattern.

- *Pair-associators* combine the stored patterns with other patterns. These networks learn to associate a pattern *B* to another pattern *A*, and so every time pattern *A* is provided to them, in complete or partial form, they provide pattern *B* as output.

- *Supervised learning* takes place when the learning process is guided by an instructor who corrects the possible errors of the network. The instructor gives the network the correct response (output) for each input pattern. According to the difference between the output calculated by the network and the correct response, the network varies the weights of the connections in order to produce an output as similar as possible to the correct responses provided by the instructor.

- *Unsupervised learning* occurs when there is no instructor and the network must learn on its own to classify the input patterns. Unsupervised networks classify input data into a set of internal categories or clusters based on the relationships that the network finds out between data, without an instructor that provides the correct output to the network. These neural networks are used when it is not known exactly how to classify the available data.

Since the first studies by McCulloch and Pitts, many models of neural networks have been proposed (Knight, 1990; Widrow and Lehr, 1990). Some of the most known neural network models are Hopfield Networks, Perceptron, Multilayer Networks, Boltzmann Machine, Self-Organizing Maps, and Adaptive Resonance Networks.

Some of these neural network models have been used to build data mining applications (Bigus, 1996). Examples of systems based on neural networks which are used to build data mining applications are Neural Connection, IBM Neural Network Utility, NeuroSolutions, BrainMaker, and ModelQuest.

A neural network learns from data provided to it in the same way people learn from their experience and on the basis of their knowledge make decisions. In fact, neural networks receive as input a set of data that may contain "dirty" information and try to reach some conclusions (output) according to what they learned in the past and that is encoded in the values of neurons and in the weights associated with their connections. Because neural networks possess the ability to recognize patterns from sets of input data, they can make a series of typical data mining operations such as classification of a dataset, clustering, association of two or more data, and the analysis of relationships between data.

As an example we briefly describe the way in which a multilayer neural network allows us to achieve a classification function. This neural network model is composed of several layers of neurons such that the output of a neuron in one layer is connected with the input of all neurons in the next layer (see Figure 1.3). The input of the first layer is the input of the neural network, and the output of the last layer is the output of the network. A multilayer network is suitable for the implementation of a classification function that assigns an object (data) defined by an input vector $[x_1, x_2, ..., x_n]$ to one or more classes $C_1, C_2, ..., C_m$ that are the output neurons of the network. If the object belongs to class C_i, the output neuron C_i has a value of 1; otherwise, if the object does not belong to class C_i, the output neuron C_i has a value of 0.

Figure 1.3 shows an example of a neural network with $n = 5$ and $m = 4$. Thus, by selecting an appropriate set of weights for all the connections of the neural network, it can implement a set of classification tasks. The correct choice of weights can be made by providing to the network a number of examples that are represented by input values and the corresponding output values (supervised learning). Under this approach, an example is

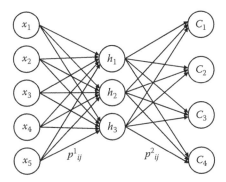

FIGURE 1.3   Example of multilayer neural network for data classification.

a vector of input values with the desired output value, where the set of C_i have the values 1 or 0 depending on whether the input belongs or does not belong to class i. The network will adjust the weights of connections based on the examples given, and at the end of the learning procedure it will be ready to classify new objects, different from those of the training set.

Based on these features, neural networks represent a natural and useful model to perform data mining in large databases. In addition, it should be noted that because of their structure, neural networks are well suited to be implemented on parallel computers (Talia, 1993) as discussed in Section 1.3.1.3. This feature allows us to use neural networks to implement high-performance data mining applications.

1.2.2.6 Genetic Algorithms

Neural networks have been proposed as a simplified model of the human brain, and genetic algorithms are a model of the biological mechanisms that cause natural selection of species. In fact, genetic algorithms are stochastic search algorithms based on the mechanics of natural selection and the comparison with the genetics observed in nature. They combine survival of the strongest elements of the population with the exchange of structured knowledge to form an innovative search algorithm. Every generation, a new set of artificial creatures (strings) is created using substrings of the best elements of the previous generation. Occasionally, new parts are tried to introduce possible improvements.

Genetic algorithms were introduced by John Holland (1975). The goal of Holland's research was twofold: to abstract and rigorously explain the adaptation process of natural systems and to design software systems that maintain the natural mechanism. This approach has led to important discoveries in the natural sciences and artificial systems. Although genetic algorithms are based on randomness, it is not a simple random search. In fact, they explore historical information effectively to establish new search points that have better features. Genetic algorithms belong to the category of probabilistic algorithms but are very different from random algorithms because they combine elements of stochastic search and guided search.

A central research theme in genetic algorithms has been the *robustness* (i.e., the right balance between efficiency and effectiveness needed to survive in many different environments). In this context, the term robustness has a different meaning from that understood in other areas of computing, such as fault tolerance and reliability of computer systems. The implications of robustness in artificial systems are manifold: if an artificial system can be made more robust, any costly redesigns can be reduced or completely eliminated. If high levels of adaptation are reached, the systems can perform their functions longer and better. The designers of both software and hardware artificial systems can be positively affected by the robustness, efficiency, and flexibility of biological systems: features such as improvement, reproduction, and care of defects are the rule in natural systems, while they appear in simple form only in the most sophisticated artificial systems.

After having established itself as a valid approach to problems that require an efficient search, genetic algorithms are finding application in engineering, science, and business. The main reasons behind this success are clear: these algorithms are computationally simple but powerful in their search for improvement, and they are not limited by restrictive assumptions about the search space (e.g., continuity and differentiability), as in the case of classic search algorithms in the optimization field.

As mentioned before, genetic algorithms perform a multidirectional search by maintaining a population of potential solutions and encouraging the formation of blocks of information and their exchange between solutions. The population goes through a simulated evolution: at each generation relatively "good" solutions reproduce, while relatively "bad" ones die. To distinguish the goodness of the solutions, a *fitness* function is used to perform the evaluation of the solutions obtained as possible solutions to the problem.

In a genetic algorithm, a *gene* is usually represented by a bit, a character (byte), or an integer, depending on the cardinality of the set of values that the gene can assume. The set of genes of an individual form the *chromosome.* The chromosome is usually represented by an array of genes, as defined above. In turn, the population is represented as an array or any collection of individuals. After specifying the initial condition defined by an initial population $P(0)$ of possible solutions, the structure of a genetic algorithm can be represented as composed of the following steps:

1. $t = t + 1$

2. Selection of $P(t)$ from $P(t - 1)$

3. Recombination of $P(t)$

4. Evaluation of $P(t)$

where t is the evolutionary clock, and $P(t)$ is the population at time t. These operations are repeated until a satisfactory solution is reached. At each iteration t, a genetic algorithm maintains a population of potential solutions, $P(t) = \{s_1^t, \ldots, s_n^t\}$. Each solution s_i^t is evaluated to measure its fitness (i.e., its goodness in the environment to which it belongs). Then, a new population (iteration $t + 1$) is constructed by selecting the best individuals. Some individuals of this new population go through reproduction and mutation to form new solutions.

Reproduction, called *crossover,* combines features of two parents to form a pair of children, swapping the corresponding segments of the parents. For example, if the parents are represented by the vectors $(a_1, b_1, c_1, d_1, e_1, f_1)$ and $(a_2, b_2, c_2, d_2, e_2, f_2)$, then crossing the chromosomes after the second gene, the children $(a_2, b_2, c_1, d_1, e_1, f_1)$ and $(a_1, b_1, c_2, d_2, e_2, f_2)$ can be generated. The intuitive justification of crossover is the exchange of information between different potential solutions. Figure 1.4 shows an example of crossover. Mutation arbitrarily alters one or more genes of the selected chromosome by using a random change with a predetermined probability (see Figure 1.5).

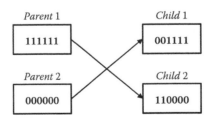

FIGURE 1.4   Example of crossover.

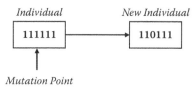

FIGURE 1.5 Example of mutation.

The intuitive justification of mutation is the introduction of variability in the population to introduce solutions that would otherwise not be explored only by crossover.

A genetic algorithm applied to a specific problem should have these components:

1. A genetic representation for potential solutions to the problem.

2. A way to create an initial population of possible solutions.

3. An evaluation function that plays the role of the environment by assessing the goodness of the solutions found.

4. Genetic operators to create and mutate their children during the recombination phase.

5. Values for the various algorithm parameters (population size, probabilities of applying the genetic operators).

The research activities carried out in the area of machine learning techniques through the use of genetic algorithms can be grouped into two main approaches:

- *Pittsburgh approach* (Smith, 1980): Any individual or population entity is a set of rules that represent a complete solution to the problem (*disjunction of the elementary descriptions*); the entire population represents a set of complete descriptions that are competing with each other. The size of each individual is then a multiple of the length of each single description. To perform the crossover operation both strings that correspond to the parent individuals are cut in the same point modulo the length of the disjunction of the descriptions. In this way, the obtained child individuals represent themselves complete solutions.

- *Michigan approach* (Holland, 1986): Each element of the population consists of individual rules that represent a single partial solution to the problem (*elementary description*); the entire population forms a single classification rule (i.e., a rule whose description is the disjunction of all the descriptions represented by the population individuals). In this case, the crossover operation is performed by breaking the strings of two individuals in the same position, and by creating two new strings by combining the first part of the string of one parent with the second part of the string of the other parent, and vice versa.

In both approaches, the goodness of an individual is measured based on its classification accuracy (ratio of positive examples/negative examples covered) and its generality (number of examples covered over the total number of examples covered by the whole population).

An example of the use of genetic algorithms to perform a data mining task is that designed to characterize the behavior of customers of a fruit and vegetable retail chain in a central area of a big city. The database contains sales data of over 100 products for a period of 50 weeks. Data include many fields such as price, quantity sold, vendor, and so forth. The database also contains weather information for each period in terms of rainfall per week, temperature, and sunshine hours (Radcliffe and Surry, 1994).

The genetic algorithm is designed to discover the best rules that find correlations between data (expressed by rules) on the products sold, as a function of the weather in the period of interest. For example, the amount of strawberries sold when the temperature is lower than the average for a given period, or the price of apples in a given period depending on the rainfall.

The genetic algorithm starts by generating random rules (each rule is realized by a string); then the rules are changed by the algorithm with mutation and crossover operators. For every new rule generated, the algorithm evaluates its fitness over the entire database. The fitness of a rule is expressed in terms of coverage and classification accuracy of the rule. Through the mutation operator, one or more components (or conditions) of the rule are changed to obtain a new rule. In our example, the product, the period of interest, or a numeric value of the rule in question can be modified. The crossover operator produces a new rule by selecting a portion of the rule (clause) from two "parent" rules. A bias determines the probability of selecting a portion of the rule, for instance, *price(mac apples, yesterday) < 1.5 price(golden apples, last week).*

After a certain number of generations, the system will present a user with the set of rules having the best fitness. The rules will contain information on the sale of products as a function of the weather conditions, for example,

1. Sale of strawberries: *If rainfall is less than 1.3 mm, then they will be sold more than 120 kg per week.*

2. Sale of golden apples in the last 6 months: *If the average hours of sunshine + 4.5 × their price in Euro per kg is greater than 25 Euros*, then *the total sales of golden apples will be greater than 1,950,000 Euros.*

One of the first machine learning systems based on genetic algorithms is LS-1 (Smith, 1984), which introduced a structured representation based on the semantics of the domain where the problem to be solved through genetic operators lies. Another system that, like LS-1, is based on the Pittsburgh approach is GABIL (Genetic Algorithm-Based Inductive Learning), in which descriptions of the concepts are defined through a collection of classification rules that may overlap. Other systems based on genetic algorithms are OMEGA Predictive Modeling System, Evolver, and Genalytics GA3.

1.3 PARALLEL KNOWLEDGE DISCOVERY

Data mining algorithms working on very large datasets often take very long times on conventional computers to get results. Sequential data mining applications often require several days or weeks to complete their task. One approach to reduce response time is sampling. But in some cases, reducing data might result in inaccurate models. In some other cases, it is not useful (e.g., outlier detection). The other viable approach is based on the use of parallel computing. In fact, high-performance computers and parallel data mining algorithms can offer an effective way to mine very large datasets and obtain results in reasonable time. Parallel computing systems can bring significant benefits in the implementation of data mining and knowledge discovery applications by means of the exploitation of inherent parallelism of data mining algorithms.

The main goals of the use of parallel computing technologies in KDD are:

- Performance improvements of existing techniques

- Implementation of new (parallel) techniques and algorithms

- Concurrent analysis using different data mining techniques in parallel and result integration to get a better model (i.e., more accurate results)

We identify three main strategies in the exploitation of parallelism in data mining algorithms:

- Independent parallelism

- Task parallelism

- Single Program Multiple Data (SPMD) parallelism

Independent parallelism is exploited when processes are executed in parallel in an independent way. As shown in Figure 1.6, each process generally accesses to the whole dataset and does not communicate or synchronize with other processes. According to task parallelism (or *control parallelism*), illustrated in Figure 1.7, each process executes different operations on (a partition of) the dataset. Finally, in Single Program Multiple Data (SPMD) parallelism, illustrated in Figure 1.8, a set of processes execute in parallel the same algorithm on different partitions of a dataset, and processes cooperate to exchange partial results. These three strategies for parallelizing data mining algorithms are not necessarily alternative. They can be combined to implement hybrid parallel data mining algorithms in order to improve both performance and accuracy of results.

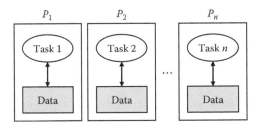

FIGURE 1.6 *Independent parallelism.* A set of tasks (either identical or different) are executed as independent processes. Generally, each process operates on the whole dataset.

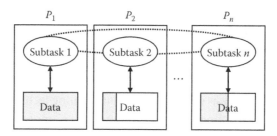

FIGURE 1.7 *Task parallelism.* The data mining task is composed of a set of different subtasks that are executed as cooperating processes. Each process operates on a subset of data or on the whole dataset.

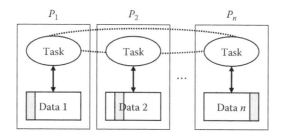

FIGURE 1.8 *Single Program Multiple Data (SPMD) parallelism.* A set of identical tasks are executed as cooperating processes. Each process operates on a different partition of the dataset.

In combination with strategies for parallelization, different data partition strategies may be used:

- *Sequential partitioning*: Separate partitions are defined without overlapping among them

- *Cover-based partitioning*: Some data can be replicated on different partitions

- *Range-based query partitioning*: Partitions are defined on the basis of some queries that select data according to attribute values.

1.3.1 Parallelism in Data Mining Techniques

This section presents different parallelization strategies for each data mining technique and describes some parallel data mining tools, algorithms, and systems. Table 1.1 contains the main data mining tasks (as introduced

TABLE 1.1 Data Mining Tasks and Used Techniques

Data Mining Tasks	Data Mining Techniques
Classification	Induction, neural networks, genetic algorithms
Association	Apriori, statistics, genetic algorithms
Clustering	Neural networks, induction, statistics
Regression	Induction, neural networks, statistics
Episode discovery	Induction, neural networks, genetic algorithms
Summarization	Induction, statistics

in Section 1.2.1), and for each task the main techniques used to solve them are listed. In the following section we describe different approaches for parallel implementation of some techniques listed in Table 1.1.

1.3.1.1 Parallel Decision Trees

Classification is the process of assigning new objects to predefined categories or classes. As described in Section 1.2.2.1, decision trees are an effective technique for classification. They are tree-shaped structures that represent sets of decisions. These decisions generate rules for the classification of a dataset. The tree leaves represent the classes, and the tree nodes represent attribute values. The path from the root to a leaf gives the features of a class in terms of attribute–value pairs.

> *Task parallel approach:* According to the task parallel approach, one process is associated to each subtree of the decision tree that is built to represent a classification model (see Figure 1.9). The search occurs in parallel in each subtree, thus the degree of parallelism P is equal to the number of active processes at a given time. A possible implementation of this approach is based on farm parallelism in which there is one master process that controls the computation and a set of P workers that are assigned to the subtrees.

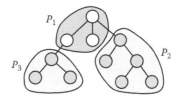

FIGURE 1.9 Task parallel approach for decision trees induction.

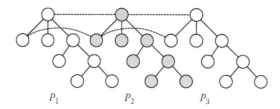

FIGURE 1.10 *Single Program Multiple Data* (SPMD) approach for decision trees induction.

SPMD approach: In the exploitation of SPMD parallelism, each process classifies the items of a subset of data (see Figure 1.10). The P processes search in parallel in the whole tree using a partition D/P of the dataset D. The global result is obtained by exchanging partial results. The dataset partitioning may be operated in two main different ways:

- By partitioning the D tuples of the dataset: D/P per processor.

- By partitioning the n attributes of each tuple: D tuples of n/P attributes per processor.

In Kufrin (1997) a parallel implementation of the C4.5 algorithm that uses the independent parallelism approach is discussed. Other significant examples of parallel algorithms that use decision trees are SPRINT (Shafer, Agrawal, and Mehta, 1996), and Top-Down Induction of Decision Trees (Pearson, 1994).

1.3.1.2 Parallel Association Rules Discovery

Association rules algorithms, such as Apriori, allow automatic discovery of complex associations in a dataset. The task is to find all frequent itemsets (i.e., to list all combinations of items that are found in a sufficient number of examples). Given a set of transactions D, as described in Section 1.2.2.4, the problem of mining association rules is to generate all association rules that have support (how often a combination occurred overall) and confidence (how often the association rule holds true in the dataset) greater than the user-specified minimum support and minimum confidence, respectively.

SPMD approach: In the SPMD strategy the dataset D is partitioned among the P processors, but candidate itemsets I are replicated on each processor. Each process p counts in parallel the partial support

S_p of the global itemsets on its local partition of the dataset of size D/P. At the end of this phase the global support S is obtained by collecting all local supports S_p. The replication of the candidate itemsets minimizes communication but does not use memory efficiently. Due to low communication overhead, scalability is good.

Task parallel approach: In this case both the dataset D and the candidate itemsets I are partitioned on each processor. Each process p counts the global support S_i of its candidate itemset I_p on the entire dataset D. After scanning its local dataset partition D/P, a process must scan all remote partitions for each iteration. The partitioning of dataset and candidate itemsets minimizes the use of memory but requires high communication overhead in distributed memory architectures. Due to communication overhead this approach is less scalable than the SPMD approach.

Hybrid approaches: Combinations of different parallelism approaches can be designed. For example, SPMD and task parallelism can be combined by defining C clusters of processors composed of the same number of processing nodes. The dataset is partitioned among the C clusters, thus each cluster is responsible to compute the partial support S_c of the candidate itemsets I according to the SPMD approach. Each processor in a cluster uses the task parallel approach to compute the support of its disjoint set of candidates I_p by scanning the dataset stored on the processors of its cluster. At the end of each iteration the clusters cooperate with each other to compute the global support S.

The Apriori algorithm is the most known algorithm for association rules discovery. Several parallel implementations have been proposed for this algorithm. In Agrawal and Shafer (1996) two different parallel algorithms called Count Distribution (CD) and Data Distribution (DD) are presented. The first is based on independent parallelism, and the second is based on task parallelism. In Han, Karypis, and Kumar (2000) two different parallel approaches to Apriori called Intelligent Data Distribution (IDD) and Hybrid Distribution (HD) are presented. A complete review of parallel algorithms for association rules can be found in Zaki (1999).

1.3.1.3 Parallel Neural Networks

Neural networks, introduced in Section 1.2.2.5, are a biology-inspired model of parallel computing that can be used in knowledge discovery.

Supervised neural networks are used to implement classification algorithms, and unsupervised neural networks are used to implement clustering algorithms.

A lot of work on parallel implementation of neural networks has been done in the past. Theoretically, each neuron can be executed in parallel, but in practice the grain of processors is generally larger than the grain of neurons. Moreover, the processor interconnection degree is restricted in comparison with neuron interconnection. Hence, a subset of neurons is generally mapped on each processor. There are several different ways to exploit parallelism in a neural network:

- *Parallelism among training sessions*: It is based on simultaneous execution of different training sessions.

- *Parallelism among training examples*: Each processor trains the same network on a subset of $1/P$ examples.

- *Layer parallelism*: Each layer of a neural network is mapped on a different processor.

- *Column parallelism*: The neurons that belong to a column are executed on a different processor.

- *Weight parallelism*: Weight summation for connections of each neuron is executed in parallel.

These parallel approaches may be combined to form different hybrid parallelization strategies. Different combinations can raise different issues to be faced for efficient implementation such as interconnection topology, mapping strategies, load balancing among the processors, and communication latency.

Typical parallelism approaches used for the implementation of neural networks on parallel architectures are task parallelism, SPMD parallelism, and farm parallelism.

A parallel data mining system based on neural networks is Clementine. Several task-parallel implementations of back-propagation networks and parallel implementations of self-organizing maps have been implemented for data mining tasks. Finally, Neural Network Utility (Bigus, 1996) is a neural network-based data mining environment that has also been implemented on an IBM SP2 parallel machine.

1.3.1.4 Parallel Genetic Algorithms

Genetic algorithms are used today for several data mining tasks such as classification, association rules, and episode discovery. Parallelism can be exploited in the following three main phases of a genetic algorithm without modifying the behavior of the algorithm in comparison to the sequential version:

- Population initialization

- Fitness computation

- Execution of the mutation operator

On the other hand, the parallel execution of selection and crossover operations requires the definition of new strategies that modify the behavior (and results) of a genetic algorithm in comparison to the sequential version. The most used approach is called *global parallelization*. It is based on the parallel execution of the fitness function and mutation operator while the other operations are executed sequentially. However, there are two possible SPMD variants:

- Each processor receives a subset of elements and evaluates their fitness using the entire dataset D.

- Each processor receives a subset D/P of the dataset and evaluates the fitness of every population element (data item) on its local subset.

Global parallelization can be effective when very large datasets are to be mined. This approach is simple and has the same behavior of its sequential version. However, its implementations did not achieve very good performance and scalability on distributed memory machines because of communication overhead.

Two different parallelization strategies that can change the behavior of the genetic algorithm are the *island model* (coarse grained), where each processor executes the genetic algorithm on a subset N/P of elements (subdemes) and periodically the best elements of a subpopulation are migrated toward the other processors, and the *diffusion model* (fine grained), where population is divided into a large number of subpopulations composed of few individuals (D/n where $n \gg P$) that evolve in parallel. Several subsets are mapped on one processor. Typically, elements are arranged in a regular topology (e.g., a grid). Each element evolves in parallel and executes the selection and crossover operations with the neighboring elements.

A simple strategy is the independent parallel execution of P independent copies of a genetic algorithm on P processors. The final result is selected as the best one among the P results. Different parameters and initial populations should be used for each copy. In this approach there is no communication overhead. The main goal here is not getting higher performance but better accuracy. Some significant examples of data mining systems based on the parallel execution of genetic algorithms are GA-MINER, G-NET, and REGAL (Neri and Giordana, 1995).

1.3.1.5 Parallel Cluster Analysis

Clustering algorithms arrange data items into several groups, called *clusters*, so that similar items fall into the same group and different items belong to different groups. This is done without any suggestion from an external supervisor, so classes are not given *a priori*, but they must be discovered by the algorithm. When used to classify large datasets, clustering algorithms are very computing demanding. Clustering algorithms can roughly be classified into two groups: hierarchical and partitioning methods. Hierarchical methods generate a hierarchical decomposition of a set of N items represented by a *dendogram*. Each level of a dendogram identifies a possible set of clusters. Dendograms can be built starting from one cluster and iteratively splitting this cluster until N clusters are obtained (*divisive methods*), or starting with N clusters and merging at each step a couple of clusters until only one is left (*agglomerative methods*).

Partitioning methods divide a set of objects into K clusters using a distance measure. Most of these approaches assume that the number K of groups has been given *a priori*. Usually these methods generate clusters by optimizing a criterion function. The K-means clustering is a well-known and effective method for many practical applications which employs the squared error criterion.

Parallelism in clustering algorithms can be exploited both in the clustering strategy and in the computation of the similarity or distance among the data items by computing on each processor the distance/similarity of a different partition of items. In the parallel implementation of clustering algorithms the three main parallel strategies described in Section 1.3 can be exploited.

Independent parallel approach: Each processor uses the whole dataset D and performs a different classification based on a different number of clusters K_p. To get the load among the processors balanced, until the clustering task is complete a new classification is assigned to a processor that completed its assigned classification.

Task parallel approach: Each processor executes a different task that composes the clustering algorithm and cooperates with other processors exchanging partial results. For example, in partitioning methods processors can work on disjoint regions of the search space using the whole dataset. In hierarchical methods a processor can be responsible for one or more clusters. It finds the nearest neighbor cluster by computing the distance among its cluster and the others. Then all the local shortest distances are exchanged to find the global shortest distance between two clusters that must be merged. The new cluster will be assigned to one of the two processors that handled the merged clusters.

SPMD approach: Each processor executes the same algorithm on a different partition D/P of the dataset to compute partial clustering results. Local results are then exchanged among all the processors to get global values on every processor. The global values are used in all processors to start the next clustering step until a convergence is reached or a given number of steps are executed. The SPMD strategy can also be used to implement clustering algorithms where each processor generates a local approximation of a model (classification), which at each iteration can be passed to the other processors that in turn use it to improve their clustering model.

In Olson (1995), a set of hierarchical clustering algorithms and an analysis of time complexity on different parallel architectures can be found. An example of parallel implementation of a clustering algorithm is P-CLUSTER (Judd, McKinley, and Jain, 1996). Other parallel algorithms are discussed in Bruynooghe (1989), Li and Fang (1989), and Foti, Lipari, Pizzuti, and Talia (2000). In particular, in Foti, Lipari, Pizzuti, and Talia (2000), an SPMD implementation of the AutoClass algorithm named *P-AutoClass* is described. That paper shows interesting performance results on distributed memory MIMD (Multiple Instruction, Multiple Data)

TABLE 1.2   Turnaround Time and Speedup
of P-AutoClass

Processors	Turnaround Time (sec)	Speedup
1	14,683	1.0
2	7,372	2.0
4	3,598	4.1
6	2,528	5.8
8	2,248	6.5
10	1,865	7.9

machines. Table 1.2 shows experimental performance results obtained by running P-AutoClass on a parallel machine using up to 10 processors for clustering a dataset composed of 100,000 tuples with two real valued attributes. In particular, Table 1.2 contains turnaround times and absolute speedup on 2, 4, 6, 8, and 10 processors. We can observe how the system behavior is scalable; speedup on 10 processors is about 8, and the turnaround time significantly decreases from 245 to 31 minutes.

1.3.1.6 Architectural and Research Issues
In presenting the different strategies for the parallel implementation of data mining techniques, we did not address architectural issues such as:

- Distributed memory versus shared memory implementation

- Interconnection topology of processors

- Optimal communication strategies

- Load balancing of parallel data mining algorithms

- Memory usage and optimization

- Input/Output (I/O) impact on algorithm performance

- Exploitation of multicore and many-core architectures

These issues (and others) must be taken into account in the parallel implementation of data mining techniques. The architectural issues are strongly related to the parallelization strategies, and there is a mutual influence between knowledge extraction strategies and architectural features. For instance, increasing the parallelism degree in some cases corresponds to an increment of the communication overhead among the processors.

However, communication costs can also be balanced by the improved knowledge that a data mining algorithm can get from parallelization. At each iteration the processors share the approximated models produced by each of them. Thus each processor executes a next iteration using its own previous work and also the knowledge produced by the other processors. This approach can improve the rate at which a data mining algorithm finds a model for data (knowledge) and can make up for lost time in communication.

Parallel execution of different data mining algorithms and techniques can be integrated to get not just high performance but also high accuracy. Here we list some promising research issues in the parallel data mining area:

- It is still necessary to develop environments and tools for interactive high-performance data mining and knowledge discovery.

- The use of parallel knowledge discovery techniques in text mining must be extensively investigated.

- Parallel and distributed Web mining is a very promising area for exploiting high-performance computing techniques.

- Integration of parallel data mining techniques with parallel databases and data warehouses is a crucial aspect for private enterprises and public organizations.

1.4 DISTRIBUTED KNOWLEDGE DISCOVERY

In many scenarios, data can be collected from many sources and stored at enormous speeds (GB/hour). Both such aspects imply that applications have to deal with massive volumes of data. As mentioned before, the analysis of large datasets requires powerful computational resources. A major issue in knowledge discovery is scalability with respect to the very large size of current-generation and next-generation databases, given the excessively long processing time taken by (sequential) data mining algorithms on realistic volumes of data. In fact, data mining algorithms working on very large datasets take a very long time on conventional computers to get results. In order to improve performance and deal with data stored on distributed repositories, distributed data mining algorithms and distributed knowledge discovery techniques have been designed. Those algorithms and techniques can integrate both sequential and parallel data mining algorithms that can run on single computing nodes or on parallel computers, which are part of a computing infrastructure including many geographically distributed sites.

Distributed knowledge discovery (DKD), or *distributed data mining* (DDM), works by analyzing data in distributed computing systems and pays particular attention to the trade-off between centralized collection and distributed analysis of data. This approach is particularly suitable for applications that typically deal with very large amounts of data (e.g., transaction data, scientific simulation and telecommunication data) which cannot be analyzed in a single site on traditional machines in acceptable times or that need to mine data sources located in remote sites, such as Web servers or departmental data owned by a large enterprise, or data streams coming from sensor networks or satellites.

Traditional warehouse-based architectures for data mining are assumed to have a centralized data repository. Such a centralized approach is fundamentally inappropriate for most of the distributed and ubiquitous data mining applications. In fact, the long response time, lack of proper use of distributed resources, and the fundamental characteristic of centralized data mining algorithms do not work well in distributed environments. A scalable solution for distributed applications calls for distributed processing of data, controlled by the available resources and human factors.

For example, let us consider an ad hoc wireless sensor network where the different sensor nodes are monitoring some time-critical events. Central collection of data from every sensor node may create traffic over the limited bandwidth wireless channels, and this may also drain a lot of power from the devices. In fact, most of the applications that deal with time-critical distributed data are likely to benefit by paying careful attention to the distributed resources for computation, storage, and the cost of communication. As another example, let us consider the Web as it contains distributed data and computing resources. An increasing number of databases (e.g., weather databases, oceanographic data, etc.) and data streams (e.g., emerging disease information, etc.) are currently made available online, and many of them change frequently. It is easy to think of many applications that require regular monitoring of these diverse and distributed sources of data.

A distributed approach to analyze this data is aimed at being more scalable, particularly when the application involves a large number of data sites. Hence, in this case we need data mining architectures that pay careful attention to the distribution of data, computing, and communication, in order to access and use them in a near-optimal fashion. Most DDM algorithms are designed upon the potential parallelism they can apply over the given distributed data. Typically, the same algorithm operates on each distributed data site concurrently, producing one local model per site. Subsequently, all local models are aggregated to produce the final model. This schema is common to several distributed data mining algorithms. Among them, ensemble learning, meta-learning, and collective data mining are the most important.

1.4.1 Ensemble Learning

Ensemble learning (Tan, Steinbach, and Kumar, 2006) aims at improving classification accuracy by aggregating predictions of multiple classifiers. An ensemble method constructs a set of base classifiers from training data and performs classification by voting (in the case of classification) or by averaging (in the case of regression) on the predictions made by each classifier. The final result is the ensemble classifier, which tends to have higher classification quality than any single classifier composing it.

Figure 1.11 shows an example of ensemble learning for data classification:

1. The input dataset is split, using a partitioning tool, into a training set and a test set.

2. The training set is given in input to n classification algorithms that run in parallel to build n independent classification models.

3. Then, a voter tool performs an ensemble classification by assigning to each instance of the test set the class predicted by the majority of the n models generated at the previous step.

Efficiency in ensemble learning is improved because the n classification algorithms can be executed in parallel on a set of distributed nodes, thus achieving significant execution time speedups. The identification of optimal ways to combine the base classifiers is a crucial point. Prominent among these are schemes called *bagging* and *boosting*. They can increase predictive performance over a single model and are general techniques that can be applied to numerical prediction problems and to classification tasks.

Bagging (called voting for classification, averaging for regression) combines the predicted classifications (prediction) from multiple models or

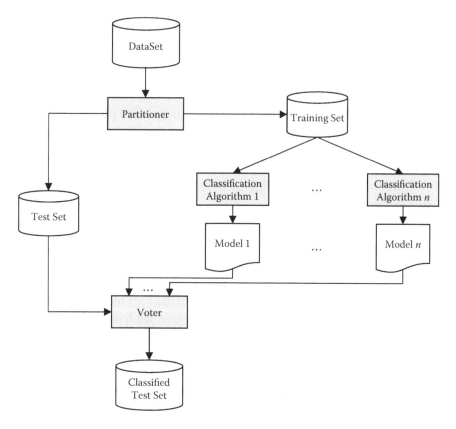

FIGURE 1.11   Example of ensemble learning for data classification.

from the same type of model for different learning data. It is also used to address the inherent instability of results when applying complex models to relatively small datasets.

Boosting also combines the decisions of different models, like bagging, by amalgamating the various outputs into a single prediction, but it derives the individual models in different ways. In bagging, the models receive equal weight, whereas in boosting, weighting is used to give more influence to the more successful ones.

1.4.2 Meta-Learning

The meta-learning technique aims at building a global model from a set of inherently distributed data sources. Meta-learning can be defined as learning from learned knowledge (Prodromidis, Chan, and Stolfo, 2000). In a data classification scenario, this is achieved by learning from

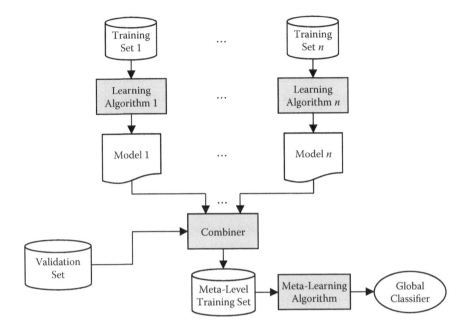

FIGURE 1.12   Building a global classifier with meta-learning.

the predictions of a set of base classifiers on a common validation set. Figure 1.12 depicts the different steps to build a global classifier from a set of distributed training sets using a meta-learning approach:

1. The initial training sets are given in input to n learning algorithms that run in parallel to build n classification models (base classifiers).

2. A meta-level training set is formed by combining the predictions of the base classifiers on a common validation set.

3. A global classifier is trained from the meta-level training set by a meta-learning algorithm.

Stacking (or *stacked generalization*) is a way of combining multiple models in meta-learning. Unlike bagging and boosting, stacking is not normally used to combine models of the same type; instead it is applied to models built by different learning algorithms. Stacking does not use a voting approach but tries to learn which classifiers are the reliable ones, using another learning algorithm (the meta-learner) to discover how best to combine the output of the base learners.

1.4.3 Collective Data Mining

Collective data mining (Kargupta et al., 2000) exploits a different strategy: instead of combining a set of complete models generated at each site on partitioned or replicated datasets, it builds the global model through the combination of partial models computed in the different sites. The global model is directly composed by summing an appropriate set of basis functions. The global function $f(x)$ that represents the global model can be expressed as

$$f(x) = \Sigma_k w_k \psi_k(x)$$

where $\psi_k(x)$ is the kth basis function, and w_k is the corresponding coefficient that must be learned locally on each site based on the stored dataset. This result is founded on the fact that any function can be expressed in a distributed fashion using a set of appropriate basis functions that may contain nonlinear terms. If the basis functions are orthonormal, the local analysis generates results that can be correctly used as components of the global model. If a nonlinear term is present in the summation function, the global model is not fully decomposable among local sites, and cross-terms involving features from different sites must be considered. As described in Kargupta, Park, Hershberger, and Johnson (2000), the main steps of this distributed data mining approach are

1. Select an appropriate orthonormal representation for the type of data model to be generated.

2. Generate at each site approximate orthonormal basis coefficients.

3. If the global function includes nonlinear terms, move a sample of datasets from each site to a central site and generate there the approximate basis coefficients corresponding to such nonlinear terms.

4. Combine the local models to generate the global model and transform it into the user-described model representation.

CHAPTER 2

Service-Oriented Computing for Data Analysis

THIS CHAPTER DISCUSSES THE PRINCIPLES of service-oriented architectures and computing, presents the basic concepts of Web services,
Grid services, and Cloud services, and discusses how service-oriented
approaches are exploited to implement knowledge discovery systems. In
particular, Section 2.1 introduces the basic concepts of service-oriented
architecture (SOA) and service-oriented computing (SOC). Section 2.2
presents the main implementations of SOA in the Web, Grid, and Cloud
contexts. Finally, Section 2.3 discusses service-oriented knowledge discovery, including some Grid-based knowledge discovery systems and testbeds.

2.1 SERVICE-ORIENTED ARCHITECTURE AND COMPUTING

A service-oriented architecture is essentially a collection of services, each
of which can communicate directly with a group of processes or functions, or may participate in a more complex business process involving
other services. Systems based on an SOA may be either used within a
single organization by a limited set of applications, or accessed via the
Internet by a large set of clients.

Services in an SOA typically satisfy a variety of fundamental enterprise requirements such as security, transactions, pooling, clustering, and

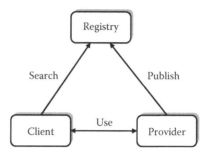

FIGURE 2.1 Actors and interactions in a service-oriented architecture (SOA).

batch processing (Papazoglou and Georgakopoulos, 2003). Each service is composed of two main parts:

- *Interface*, which describes data types, operations, transport-protocol binding, and network location of the service.

- *Implementation*, which implements the service functionalities according to the interface specifications.

An SOA is based on the interaction among three actors (see Figure 2.1):

- *Provider*, which is the server that provides the service.

- *Registry* (or *broker*), which allows clients to locate the service of interest.

- *Client* (or *requestor*), which requests and uses the service.

The following operations can be performed by the SOA actors:

- *Publication*: In order to make a service accessible, a service descriptor must be published by the registry to tell clients how to use it.

- *Search*: The client may retrieve a service descriptor either directly or by searching a registry.

- *Use*: The client invokes the service hosted by a provider, as specified by the service descriptor.

Therefore, SOA defines "who does what" in a series of interactions in which the service plays the central role. The three actors can be distributed and implemented using different technologies; the only requirement is that they use the same transmission medium.

The SOA paradigm is very effective to support the development of distributed and heterogeneous applications. In fact, SOA uses services as building blocks for the composition of distributed applications, promoting interoperability through the decoupling of service interfaces from their implementations for providing seamless and automatic connections from one software application to another. Hence, the focus is moved from technological aspects to service aspects: software applications, as well as platforms that support them or their components, are replaced by modular functions exposed as services, regardless of where they are located, and only looking at how to access and retrieve results. The SOA paradigm may also be thought of as a derivation of the object-oriented paradigm, which is widely used in software engineering, but seen from a business perspective rather than from a technology point of view.

The SOA model is at the core of service-oriented computing (i.e., the use of services to support the development of rapid, low-cost, interoperable, evolvable, and massively distributed applications) (Papazoglou et al., 2007). The main goal of SOC is to assemble application components into a loosely coupled network of services to create dynamic business processes and agile applications that span organizations and computing platforms.

Service composition is at the core of SOC. It allows the aggregation of multiple services into a single composite service that can be offered as a complete application and solution to service clients, or used as basic service in further service compositions. After performing service composition, service aggregators become service providers by publishing the service descriptions of the composite service they have created.

The terms *orchestration* and *choreography* are widely used to describe the interaction protocols that coordinate and control collaborating services (Peltz, 2003):

- *Orchestration* refers to an executable business process that may interact with both internal and external Web services. Orchestration describes how Web services can interact at the message level, including the business logic and execution order of the interactions. These interactions may span applications and organizations, resulting in a long-lived transactional process. The process is always controlled from the perspective of one of the business parties. Orchestration can be achieved using the Web Services Business Process Execution Language (WS-BPEL),* or other eXtensible Markup Language (XML)–based process standard definition languages.

- *Choreography* is associated with the public message exchanges, rules of interaction, and agreements that occur between multiple Web services, rather than a specific business process executed by a single party. Therefore, it differs from orchestration being more collaborative in nature, because each party involved in the process describes the part it plays in the interaction. Choreography tracks the sequence of messages involving multiple parties, where no one party "owns" the conversation. Choreography can be achieved using the Web Services Choreography Description Language (WS-CDL)† that specifies the common observable behavior of all participants.

An important aspect related to service composition is service customization. Supporting users through services requires the use of a wide variety of information (*profiles*) related to each user. By sharing this information across disparate systems, it is possible to simplify the use and customization of services. The primary objectives of service customization are:

- Supporting user profiles; "enter once, use everywhere."

- Sharing such profiles, even in different domains.

- Making it possible to temporarily change user preferences.

- Supporting different levels of privacy.

- Having the ability to reason about profiles by rules, using reasonably simple formalisms.

- Having the opportunity to develop personal agents that, in exploiting the user profiles and updating them continuously, may discover, invoke, and combine services on behalf of the users.

The sharing of user profiles can be achieved through a federation of heterogeneous independent systems by establishing a global schema that serves as a conceptual access point to the shared data. The federation of

---

* OASIS, 2007, April 11, Web Services Business Process Execution Language Version 2.0, http://docs.oasis-open.org/wsbpel/2.0/OS/wsbpel-v2.0-OS.html (accessed June 2012).

† W3C®, 2005, November, Web Services Choreography Description Language Version 1.0, http://www.w3.org/TR/ws-cdl-10 (accessed June 2012).

user profile data can greatly improve service customization. This can be achieved through the application of *policies* based on the profiles, which guide the choices made by the application logic of services. There are two main modes of application of policies:

- *Data-driven application* exists where there is a structure that keeps information about user preferences, and the application logic of the different services interprets the content according to their needs; this structure may evolve over time, resulting in dynamic behaviors.

- *Rule-based application* exists where families of rules are specified and then enforced by an execution engine during the use of services.

There are a variety of approaches to the use of rule-based policies (Hull et al., 2002). In general, the rule-based approach is more flexible than the data-driven one, because in practical applications rule-based languages have greater expressive power as compared to programs interpreting values.

2.2 INTERNET SERVICES: WEB, GRIDS, AND CLOUDS

SOA-based approaches are being extensively adopted by system designers to implement large-scale applications that exploit distributed services available through the Internet. Internet-enabled services have been widely proven effective as a means to integrate geographically distributed software and hardware resources into large-scale applications, both in science and business scenarios. Here, we focus on Internet-based services by discussing the most important SOC implementation (i.e., Web services, Grid services, and Cloud services). Web services provide the basis for the design and execution of business processes that are distributed over the network and available through standard Internet protocols. Grid services enable the distributed execution of long-running scientific computations using standardized mechanisms that are mostly based on stateful service interfaces. Cloud services enable on-demand network access to a large pool of computing resources (e.g., networks, servers, storage, applications) that can be dynamically provisioned to end users with minimal management effort. There are commonalities between Web, Grid, and Cloud services, with Web services basically playing the role of core technology also in Grid and Cloud computing systems. However, the difference in application focus led Grid and Cloud designers to define some specific service mechanisms that are oriented at improving either performance or resource management. The remainder of this section is organized as follows: Section 2.2.1 describes the Web service principles and technologies; Section 2.2.2 introduces the Grid computing paradigm and the Grid service standards; and finally, Section 2.2.3 discusses the role of services in Cloud computing systems.

2.2.1 Web Services

Web services are software services that can be described, located, and used through Internet protocols and by means of XML formalisms. Information on how to access services, as well as information on their semantics, is typically used by other software systems. Therefore, interactions with Web services take place through an exchange of XML messages, by using one or more standard Internet protocols. The use of XML as the base language for this technology provides a common syntax for expressing structured data in a form readable by humans, and also a schema for data sharing independent from both platform and programming language. The choice of working through common protocols used on the Internet is mainly due to the enormous spread of its standards, toward which skills and technical knowledge of developers are consolidated. Moreover, the use of common Internet protocols gives the opportunity to reuse existing hardware and software infrastructures.

Web services allow us to overcome many limitations of traditional distributed architectures based on remote components (such as Java Remote Method Invocation [RMI], Common Object Request Broker Architecture [CORBA], and Distributed Component Object Model [DCOM]). First, such architectures perform a low-level binary communication; thus, the rules for encoding and decoding transmitted data totally depend on the specific technology used. Moreover, Web services enable and encourage applications with a greater granularity, being designed to expose coarse-grained services, while traditional distributed architectures lend themselves to being used to implement fine-grained components. This means that Web services expose functionalities to a higher level of abstraction, while remote components can expose low-level invocations that are more related to implementation aspects.

Therefore, the use of Web services enables the creation of a distributed environment in which all applications and application components can interact within properly organized units, regardless of platforms and languages used, and based on a widespread and relatively simple common communication protocol. This promotes integration of applications, information, and existing processes to optimize and streamline the applications in order to improve their overall efficiency. In some areas, integration can result in a strategic competitive factor, in particular when the information system is heterogeneous and with a high level of complexity.

Web services are particularly well suited in the development of electronic transactions. For example, inside a sales network, Web services allow a company branch to check availability and place orders for products, while Web services offered by suppliers can be used to order raw materials or compare prices of different suppliers. The real competitive advantage that can be obtained is related to the greater ease and speed of access to information and processes, thus providing a wider and more detailed view of the business and a greater analysis capability.

The adoption of Web services has significant impact on the business models of organizations. The availability of an effective and standard technology reduces the cost of solutions, which then become interesting also for companies that lack a substantial budget to invest. Also the market of outsourcing services gains new capabilities with Web services. On one hand, Web services are a means of integration; on the other hand, they allow the decomposition of applications, thus facilitating organizations in deciding which services to outsource, maintaining the critical ones inside.

As defined by the W3C,* the basic requirements of a reference architecture for Web services are as follows:

- *Interoperability*: The architecture *should* enable the development of interoperable Web services across a wide set of environments, ensuring independence from underlying systems and programming languages.

- *Reliability*: The architecture *must* ensure that Web services are reliable, stable, and evolvable over time.

- *Web-friendliness*: The architecture *must* be consistent with the current and future evolution of the Web.

- *Security*: The architecture *must* provide a secure environment for online processes, addressing the security of Web services and enabling privacy protection for consumers across distributed domains and platforms.

- *Scalability and extensibility*: The architecture *must* enable scalable and extensible implementations. Web services must be modular to make possible a reduction of complexity through decomposition.

---

* W3C, http://www.w3.org (accessed June 2012).

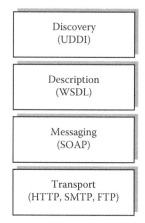

FIGURE 2.2   Web services protocol stack.

2.2.1.1 Protocol Stack

Web services are supported by a core set of protocols, each one positioned at a different layer in the protocol stack shown in Figure 2.2.

- *Transport* layer, which includes protocols for transporting messages between services and applications.

- *Messaging* layer, which is responsible for encoding messages using a standard XML format.

- *Description* layer, which is responsible for describing the interface of Web services using an XML-based language.

- *Discovery* layer, which provides the means to publish information about services into a registry, and to search such a registry to find the services of interest.

At the transport layer, the most used protocol is *HTTP* (Hypertext Transfer Protocol), but other protocols such as SMTP (Simple Mail Transfer Protocol) and FTP (File Transfer Protocol) can be used. The affirmation of HTTP as the reference protocol for the transport layer is due to three main reasons: (1) it is a stable and mature protocol; (2) existing information systems support it; (3) network firewalls often allow the passage of

HTTP packets, and thus we are not forced to intervene on the security policies to use it.

At the messaging layer, the *SOAP* protocol* specifies how to exchange information between services and applications. SOAP relies on XML for its message format and on a transport layer protocol (typically HTTP, as said before) for message transmission. The SOAP protocol consists of three parts: an envelope that defines what is in a message and how to process it, a set of encoding rules for expressing instances of application-defined data types, and a convention for representing remote procedure calls and responses.

The description layer includes the *Web Service Description Language* (*WSDL*),† an XML-based standard used to describe the interface of Web services. WSDL allows for description of (1) the operations performed by a service and the input and output messages to be exchanged for each operation; (2) service goals and obtainable results (typically using terms and concepts defined by an ontology); (3) the expected behavior of the service; (4) quality-of-service information, such as performance, reliability, scalability, security supports, and costs; and (5) procedures for the physical exchange of messages.

Finally, the discovery layer includes the *Universal Description, Discovery, and Integration* (UDDI) protocol,‡ which was designed to support publication and discovery of service descriptors. Descriptors are published in a registry, which associates information about service providers with service descriptors and organizes such descriptors into taxonomies. Each service descriptor contains a textual description of a service and a pointer to the WSDL definition of the service interface. UDDI defines a discovery protocol that allows requestors to query the registry to find the service of interest, searching either by name, keyword, or category.

Therefore, in a typical scenario of use, a client application retrieves the WSDL description of a Web service through an UDDI registry and sends to it, via HTTP, a request encoded using SOAP (see Figure 2.3).

In the remainder of this section we focus on the messaging and description layers, which are the core of the Web service protocol stack, by describing in more detail the SOAP and WSDL standards.

---

* W3C, Latest SOAP Versions, http://www.w3.org/TR/soap (accessed June 2012).

† W3C, 2001, March 15, Web Services Description Language (WSDL) 1.1, http://www.w3.org/TR/wsdl (accessed June 2012).

‡ UDDI XML.org, http://uddi.xml.org (accessed June 2012).

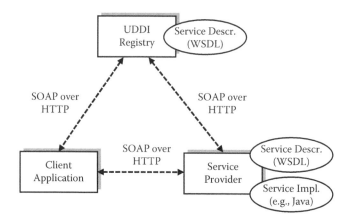

FIGURE 2.3   Typical scenario of use of a Web service.

2.2.1.2 Messaging Layer: SOAP

SOAP is the standard protocol for exchanging information between requestor and provider of a Web service. It is designed to provide a lightweight structure for the request and response messages involved in the execution of remote procedures, and for the exchange of structured data between applications.

A SOAP message contains headers associated with the transport protocols and an XML representation of the transmitted data that must follow a specific set of encoding rules. SOAP is stateless; therefore, it is not essential that sender and receiver maintain a stateful session to communicate, because all the needed state elements are contained in the message.

Applications that interact via SOAP follow, in general, a conceptual flow composed of four steps (see Figure 2.4):

1. The client application creates, through a proxy typically consisting of libraries of the language used for the application, a SOAP message to request the invocation of the service to the provider. This message is then passed to the network infrastructure, together with the address of the service provider.

2. Through the network, the SOAP message is delivered to the provider, where it is converted into objects of the language used for the service implementation, following the encoding rules specified by the message.

3. The service processes the request and generates another SOAP message as a response. This message, together with the address of the

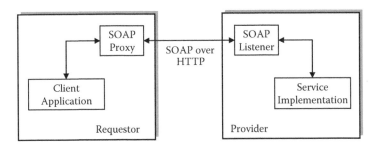

FIGURE 2.4   Interaction between service requestor and provider via SOAP.

requestor, is then passed back to the network infrastructure for delivery.

4. The SOAP response is received by the requestor and converted into objects, as specified by the encoding rules in the message. In this way, the service results are made available to the client application.

Communication with SOAP takes place by exchanging XML messages with a standard structure, which includes mandatory and optional elements. In particular, a SOAP message consists of an *Envelope* element, which contains an optional *Header* element, and a mandatory *Body* element. The *Header*, if present, can be used to include information about the type of interaction, or to transport general information about the message. The *Body* contains data intended for the ultimate message recipient, together with rules specifying their format; it may also contain an optional subelement, *Fault*, to report possible errors that occurred during the interaction.

As an example, let us consider the interaction between a requestor and a provider with the goal of using a service that returns the last yearly income of a person given his or her tax code. The provider exposes, at the uniform resource name (URN) *revenue-Income*, an operation *getLastYearlyIncome* that receives a string representing the *taxcode* of a person, and returns a floating point number representing his or her last yearly income. A possible SOAP invocation could be as follows:

```
<?xml version='1.0' encoding='UTF-8'?>
<SOAP-ENV:Envelope
   xmlns:SOAP-ENV="http://schemas.xmlsoap.org/soap/envelope/"
   xmlns:xsi="http://www.w3.org/1999/XMLSchema-instance"
   xmlns:xsd="http://www.w3.org/1999/XMLSchema">
    <SOAP-ENV:Body>
      <ns1:getLastYearlyIncome xmlns:ns1="urn:revenue-Income"
         SOAP-ENV:encodingStyle=
```

```
      "http://schemas.xmlsoap.org/soap/encoding/">
        <taxcode xsi:type="xsd:string">P3575</taxcode>
      </ns1:getLastYearlyIncome>
    </SOAP-ENV:Body>
</SOAP-ENV:Envelope>
```

The input parameter type is chosen among the scalar types of XML Schema (referred through namespaces *xsi* and *xsd*). The SOAP encoding also permits the use of complex types, such as array (with elements identified by position) and structs (by name), and of XML structures (literal encoding). The response message returned by the provider could be as follows:

```
<?xml version='1.0' encoding='UTF-8'?>
<SOAP-ENV:Envelope
  xmlns:SOAP-ENV="http://schemas.xmlsoap.org/soap/envelope/"
  xmlns:xsi="http://www.w3.org/1999/XMLSchema-instance"
  xmlns:xsd="http://www.w3.org/1999/XMLSchema">
    <SOAP-ENV:Body>
      <ns1:getLastYearlyIncomeResponse xmlns:ns1="urn:revenue-Income"
          SOAP-ENV:encodingStyle=
      "http://schemas.xmlsoap.org/soap/encoding/">
        <return xsi:type="xsd:float">56500.0</return>
      </ns1:getLastYearlyIncomeResponse>
    </SOAP-ENV:Body>
</SOAP-ENV:Envelope>
```

2.2.1.3 Description Layer: WSDL

Using service descriptions, providers can expose the information needed to invoke the service operations. The presence of the service descriptor contributes, together with the messaging protocol at the lower layer, to decoupling the service requestor by the service provider, because it is the only shared knowledge necessary to carry out the service. As mentioned earlier, the standard language used to describe Web services is WSDL.

A WSDL document specifies parameters and constraints for interacting with a given Web service, such as the format of input/output data structures, supported operations, and so on. The Web service must comply with the rules of its WSDL file. In most cases, WSDL is used in association with SOAP; however, WSDL can be used in association with different protocols that can be specified through a *binding* section, as detailed below. A WSDL document contains the following elements:

- *definitions*: This is the root element of the WSDL document; it defines the service name, declares the needed namespaces, and contains the elements listed below.

- *types*: This element describes the types of data used in the interactions, using XML Schema as the default for their representation.

- *message*: This describes a request or response message, including the message name and its input/output parameters.

- *portType*: This element combines multiple messages within an operation (e.g., a request and a response message); typically, a portType defines multiple operations.

- *binding*: This specifies the transport modes for the service messages (e.g., it can contain specific information for the use of SOAP).

- *service*: This defines the service address, typically by means of an URL.

WSDL supports four types of interactions, defined within a portType:

- *one-way*: A message from the requestor to the provider.

- *request/response*: A message from the requestor to the provider, and a response.

- *notification*: A message from the provider to the requestor.

- *solicit/response*: A message from the provider to the requestor, and a response.

As an example, let us consider the following WSDL description of an "echoing" service:

```xml
<?xml version="1.0" encoding="UTF-8"?>
<definitions name="EchoService"
    targetNamespace="http://www.example.com/wsdl/EchoService.wsdl"
    xmlns="http://schemas.xmlsoap.org/wsdl/"
    xmlns:soap="http://schemas.xmlsoap.org/wsdl/soap/"
    xmlns:tns=" http://www.example.com/wsdl/EchoService.wsdl"
    xmlns:xsd="http://www.w3.org/2001/XMLSchema">

<message name="EchoRequest">
    <part name="phrase" type="xsd:string"/>
</message>
<message name="EchoResponse">
    <part name="return" type="xsd:string"/>
</message>

<portType name="Echo_PortType">
    <operation name="echo">
        <input message="tns:EchoRequest"/>
        <output message="tns:EchoResponse"/>
```

```
      </operation>
  </portType>

  <binding name="Echo_Binding" type="tns:Echo_PortType">
     <soap:binding style="rpc"
        transport="http://schemas.xmlsoap.org/soap/http"/>
     <operation name="echo">
        <soap:operation soapAction="echo"/>
        <input>
           <soap:body
              encodingStyle="http://schemas.xmlsoap.org/soap/encoding/"
              namespace="urn:example:echoservice"
              use="encoded"/>
        </input>
        <output>
           <soap:body
              encodingStyle="http://schemas.xmlsoap.org/soap/encoding/"
              namespace="urn:example:echoservice"
              use="encoded"/>
        </output>
     </operation>
  </binding>

  <service name="Echo_Service">
     <port binding="tns:Echo_Binding" name="Echo_Port">
        <soap:address
           location="http://test.example.com:8080/soap/servlet/rpcrouter"/>
     </port>
  </service>
</definitions>
```

The description defines two message elements: one for requests and one for responses. Each message contains a *part* element: for requests, it specifies an input parameter; for responses, it specifies the return value. The *portType* element defines a single operation, called *echo*, which consists of one input message (*EchoRequest*) and one output message (*EchoResponse*).

The *binding* element refers to the portType *tns:Echo_PortType* defined previously, by providing specific details on how the *echo* operation will be transmitted. The element indicates that the binding will be performed using SOAP. The *style* attribute specifies the format used for the SOAP messages. With the format RPC, the body of the request includes an XML wrapper for the name of the invoked method and for parameters; the same applies to the response. The *transport* attribute indicates that the HTTP protocol will be used. The *operation* element indicates the specific implementation of SOAP used, specifying that the HTTP header *SOAPAction* will be used to identify the service. The *body* element specifies the details of the input and output messages; in this case, it also specifies the encoding style and the URN associated with the service. Finally, the *service* element specifies the service address, using an *address* element.

2.2.2 Grid Services

The Grid computing paradigm is today broadly used in many scientific and engineering application fields, and has registered a growing interest from business and industry for its capability to provide resource access and computing power delivery to large organizations. Grids provide coordinated access to widespread resources, including computers, data repositories, sensors, and scientific instruments, for enabling the development of innovative classes of distributed systems and applications.

The Grid computing model was designed as an effective paradigm for harnessing the power of many distributed computing resources to solve problems requiring a large number of processing cycles and involving a huge amount of data. The term *computational Grid* was adopted in the mid 1990s to denote a proposed distributed computing infrastructure for advanced science and engineering. More recently, the computational Grid model has evolved to include many kinds of advanced distributed computing fields, including commercial and business applications.

The term *Grid* comes from an analogy to the "electric power grid." As the electric power grid provides universal and standardized access to electric power to individuals and industries, the goal of computational Grids is to provide users and applications ubiquitous, standardized, and reliable access to computing power and resources. Foster and Kesselman proposed a first formal definition: "a computational Grid is a hardware and software *infrastructure* that provides *dependable, consistent, pervasive,* and *inexpensive* access to high-end computational capabilities" (1998, 18). This definition refers to an infrastructure because a Grid is mainly concerned with large-scale pooling of heterogeneous resources, including processors, sensors, data, and people. This requires a hardware infrastructure to achieve the necessary interconnections among resources and a significant software infrastructure to control the resulting ensemble.

Some years later, Foster, Kesselman, and Tuecke refined the previous definition stating that Grid computing is concerned with "coordinated *resource sharing* and problem solving in dynamic, multi-institutional *virtual organizations*" (2001, 200). Resource sharing concerns primarily direct access to computers, software, data, and other resources, as is required by a variety of collaborative problem-solving strategies in science, engineering, and industry. This sharing requires the definition of sharing rules that specify clearly what is shared, who is allowed to share, and the conditions under which sharing occurs. A set of individuals and institutions defined by such sharing rules forms a *virtual organization*.

Basic Grid protocols and services are provided by toolkits and environments such as Globus Toolkit,* Unicore,† and Condor.‡ In particular, the Globus Toolkit is the most widely used middleware in scientific and data-intensive Grid applications, and represents a de facto standard for implementing Grid systems. The toolkit addresses security, information discovery, resource and data management, communication, fault detection, and portability issues.

Ever more, Grid applications address collaboration, data sharing, cycle sharing, and other modes of interaction that involve distributed resources and services (Foster et al., 2002). The need for integration and interoperability among this increasing number of applications has led the Grid community to the design of the *Open Grid Services Architecture* (OGSA), which offers an extensible set of services that virtual organizations can aggregate in various ways (Talia, 2002).

OGSA aligns Grid technologies with Web services technologies to take advantage of important Web services properties, such as service description and discovery, automatic generation of client and service code from service description, compatibility with emerging higher-level open standards and tools, and broad commercial support (Foster et al., 2003). To achieve this goal, OGSA defines a uniform exposed service semantic, the *Grid service*, based on principles from both Grid computing and Web services technologies.

OGSA defines standard mechanisms for creating, naming, and discovering transient Grid service instances. OGSA provides location transparency and multiple protocol bindings for service instances; it supports integration with underlying native platform facilities. OGSA also defines mechanisms required for creating and composing sophisticated distributed systems, including lifetime management and notification. The result is a distributed service-oriented system based on standard mechanisms which supports the creation of complex distributed applications in scientific and business scenarios.

---

* Globus, Welcome to the Globus Toolkit Homepage, http://www.globus.org/toolkit (accessed June 2012).
† UNICORE, http://www.unicore.org (accessed June 2012).
‡ Condor: High Throughput Computing, http://www.cs.wisc.edu/condor (accessed June 2012).

The goal of this section is to introduce OGSA by describing its main concepts and features. Moreover, we introduce the *Web Services Resource Framework* (WSRF), a family of Web service standards that allow Grid programmers to implement OGSA-compliant applications.

2.2.2.1 Open Grid Services Architecture

The main goal of the *Open Grid Services Architecture* (OGSA) is to provide a well-defined set of basic interfaces for the development of interoperable Grid systems and applications. The attribute "open" is used to communicate architecture extensibility, vendor neutrality, and commitment to a community standardization process (Foster et al., 2003). OGSA adopts Web services as basic technology. As discussed above, Web services are an important paradigm focusing on simple, Internet-based standards, such as SOAP and WSDL, to address heterogeneous distributed computing.

In OGSA every resource (e.g., computer, storage, program) is represented by a service (i.e., a network-enabled entity that provides some capability through the exchange of messages). More specifically, OGSA represents everything as a *Grid service*—a Web service that conforms to a set of conventions and supports standard interfaces. Whereas Web services address only persistent service discovery and invocation, OGSA supports also transient service instances created and destroyed dynamically. Hence, a Grid service is a potentially transient Web service based on Grid protocols expressed by WSDL.

Grid services are characterized by the capabilities they offer. A Grid service implements one or more interfaces (corresponding to WSDL *portTypes*), where each interface defines a set of operations that are invoked by exchanging a defined sequence of messages.

Grid services can maintain an internal state; the existence of state distinguishes one instance of a service from another that provides the same interface. The term *Grid service instance* is used to refer to a particular instantiation of a Grid service. Grid service instances can be created and destroyed dynamically. Services may be destroyed explicitly or may be destroyed or become inaccessible as a result of some system failures such as an operating system crash or network partition.

Each Grid service instance is assigned a globally unique name, the *Grid Service Handle* (GSH), which distinguishes a specific Grid service instance from all other Grid services. The GSH carries no protocol- or instance-specific information such as network address and supported protocol bindings. This information is encapsulated into the *Grid Service*

Reference (GSR). Unlike a GSH, which is invariant, the GSR(s) for a Grid service instance can change over the lifetime of that service. A GSR has an explicit expiration time, or may become invalid at any time during the lifetime of a service, and OGSA defines appropriate mapping mechanisms for obtaining an updated GSR.

2.2.2.2 Web Services Resource Framework

The *Web Services Resource Framework* (*WSRF*) is a family of technical specifications, designed under the guidance of the Open Grid Forum, which allows Grid programmers to implement OGSA-compliant applications (Czajkowski et al., 2004).

Basically, WSRF is concerned with the creation, addressing, inspection, and lifetime management of *stateful resources*. The framework provides the means to express state as stateful resources and codifies the relationship between Web services and stateful resources in terms of the "implied resource pattern" that is a set of conventions on Web services technologies, in particular XML, WSDL, and WS-Addressing. The composition of a stateful resource and a Web service that participates in the implied resource pattern is called *WS-Resource*. The framework describes the WS-Resource definition and association with the description of a Web service interface, and describes how to make the properties of a WS-Resource accessible through a Web service interface.

WSRF is a family of five separate technical specifications: WS-Resource Lifetime, WS-Resource Properties, WS-Renewable References, WS-Service Group, and WS-Base Faults.

WS-Resource Lifetime defines mechanisms for important aspects of the WS-Resource life cycle, such as creation and destruction. WSRF does not define the message exchanges used to request creation of new WS-Resources. It simply notes that new WS-Resources may be created by some out-of-band mechanism, or by means of a pattern called a *WS-Resource factory*. A WS-Resource factory is any Web service capable of bringing one or more WS-Resources into existence. The response message of a WS-Resource factory operation typically contains at least one endpoint reference (ER) that refers to the new WS-Resource. The WS-Resource Lifetime defines two ways of destroying a WS-Resource: immediate and scheduled. This allows designers flexibility to design how their Web service applications can clean up resources no longer needed.

WS-Resource Properties defines the types and values of those components of a WS-Resource state that can be viewed and modified by service requestors through a Web service interface. The WS-Resource state is expressed as an XML *resource property document* defined using XML Schema. Service requestors may determine a WS-Resource type by retrieving the XML Schema definition via standard means. Service requestors may use Web services message exchanges to read, modify, and query the XML document representing the WS-Resource state. The term *resource property* is used to refer to an individual component of a WS-Resource state.

WS-Renewable References defines mechanisms that can be used to retrieve an updated version of an endpoint reference when it becomes invalid. A WS-Addressing endpoint reference may contain not only addressing but also policy information concerning interactions with the service. An endpoint reference made available to a client represents a copy of that policy information that may, at some point, become incoherent due to changes introduced by the authoritative source. In such situations, it becomes important to be able to renew the endpoint reference.

WS-Service Group defines a means by which Web services and WS-Resources can be aggregated or grouped together for a domain-specific purpose. In order for requestors to form meaningful queries against the contents of the service group, membership in the group must be constrained in some fashion. The constraints for membership are expressed by intension using a classification mechanism. Moreover, the members of each intension must share a common set of information over which queries can be expressed.

WS-Base Faults defines an XML Schema type for a base fault, along with rules for how this fault type is used by Web services. Even though there is nothing specific to WS-Resources in this specification, it is nonetheless used by all of the other WSRF specifications to bring consistency to the faults returned by the operations, including consistent reporting of faults relating to WS-Resource definition and use.

A separate family of specifications, called *WS-Notification*, defines a standard Web services approach to notification using a topic-based publish-subscribe pattern. WS-Notification includes three specifications:

- *WS-Base Notification* defines the Web service interfaces for notification producers and notification consumers. It includes standard message exchanges to be implemented by service providers that wish to act in these roles, along with operational requirements expected of them. This is the base specification on which the other WS-Notification specifications depend.

- *WS-Brokered Notification* defines the Web services interface for notification brokers (i.e., intermediaries that allow message publication from entities that are not themselves service providers).

- *WS-Topics* defines a mechanism to organize and categorize items of interest for subscription known as "topics." WS-Topics defines topic expression dialects that can be used as subscription expressions in subscribe request messages and other parts of the WS-Notification system.

2.2.3 Cloud Services

Cloud computing is a model of computing in which dynamically scalable and often virtualized resources are provided as services over the Internet. Users are not required to have knowledge of, expertise in, or control over the technology infrastructure in the "Cloud" that supports them. A number of features define Cloud applications, services, data, and infrastructure:

- *Remotely hosted*: Services and data are hosted on a remote infrastructure.

- *Ubiquitous*: Services or data are available from anywhere.

- *Pay-per-use*: The result is a utility computing model similar to that of traditional utilities, like gas and electricity, where you pay for what you want.

More formally, we can use the NIST (National Institute of Standards and Technology) definition of Cloud computing to introduce its main characteristics: "Cloud computing is a model for enabling convenient, on-demand network access to a shared pool of configurable computing resources (e.g., networks, servers, storage, applications, and services) that can be rapidly provisioned and released with minimal management effort or service provider interaction".* From this definition we can identify the main features of Cloud computing systems, which are on-demand self-service, broad network access, resource pooling, rapid elasticity, and measured service. In deploying Cloud infrastructures different approaches can be used:

---

* National Institute of Standards and Technology, http://csrc.nist.gov/publications/nistpubs/ 800-145/SP800-145.pdf (pg. 2) (accessed June 2012).

- *Private Cloud*, an enterprise- or organization-owned or -leased infrastructure that is not offered to users outside of the enterprise/organization.

- *Community Cloud*, a shared infrastructure for a specific community composed by many users.

- *Public Cloud*, high-performance large infrastructure sold to the public for different uses.

- *Hybrid Cloud*, a composition of two or more Clouds mentioned above.

As mentioned before, Cloud computing extends the existing trend of making services available over the Internet. In particular, Cloud infrastructures have adopted the Web services paradigm for delivering its capabilities. Everybody has recognized the value of Web-based interfaces to their applications, whether they are made available to users over the network, or whether they are internal applications that are made available to authorized users, partners, suppliers, and consultants.

Cloud service providers offer their services according to three main categories: *Infrastructure as a Service*, *Platform as a Service*, and *Software as a Service*. Infrastructure as a Service (*IaaS*) delivers storage and compute capabilities as standardized services over the network. Central processing units (CPUs), servers, storage devices, switches, routers, and other hardware/software systems are pooled and made available to applications and users. Amazon EC2 is an example of an IaaS Cloud. Platform as a Service (*PaaS*) encapsulates software and provides it as a service that can be used to build higher-level services or complete applications (e.g., .NET). PaaS can provide single features or complete operations of operating systems, middleware, development environments, and application software. The Software as a Service (*SaaS*) category features a complete application offered as a service on demand (e.g., Google Docs). A single instance of the software runs on the Cloud and services multiple end users or client organizations.

2.3 SERVICE-ORIENTED KNOWLEDGE DISCOVERY

To access and analyze massive data volumes, single high-performance computers can be inadequate, and they have to be networked into Grids or networks in order to divide and conquer a large number of computationally and data-intensive tasks. As discussed before, Cloud computing has joined this scenario and allows for the distribution of data and computing load to a large number of computers accessed through a service-oriented inter-face. The need to harness multiple distributed computers for a given com-plex application such as a knowledge discovery process, gave rise to novel software paradigms, foremost of which is service-oriented computing.

Service-oriented knowledge discovery exploits service-oriented archi-tectures with functionalities for identifying, accessing, and orchestrating local and remote data/information resources and mining tools into task-specific distributed workflows. The major challenge here is the integra-tion of these distributed and heterogeneous resources and software into a coherent and effective knowledge discovery process that implements a scalable, extensible, interoperable, modular, and easy-to-use knowledge discovery system that can run on Web, Grid, or Cloud sites accessible over the Internet according to the service-oriented paradigm.

This programming approach results in service-oriented architectures for knowledge discovery that relies on Web services to achieve scalability, extensibility, and interoperability. It offers simple abstractions for users, platform independence, and supports computationally intensive process-ing on very large amounts of data.

In the next section we analyze service-oriented knowledge discovery on Grids. Here we mention some knowledge discovery systems based on the SOA model that have been developed for the Web. Some examples of such systems are Open DMIX (Grossman et al., 2004), FAEHIM (Shaikh Ali, Rana, and Taylor, 2005), Anteater (Guedes, Meira, and Ferreira, 2006), and SOMiner (Birant, 2011).

2.3.1 Grid-Based Knowledge Discovery

As mentioned in Section 2.2.2, the Grid emerged as an efficient comput-ing infrastructure to develop applications over geographically distributed sites, providing for protocols and services enabling the integrated and seamless use of remote computing power, storage, software, and data, managed and shared by different organizations.

A wide set of applications are being developed for the exploitation of Grid platforms. Since application areas range from scientific computing to industry and business, specialized services are required to meet needs in different application contexts. In particular, *data Grids* have been designed to easily store, move, and manage large datasets in distributed data-intensive applications.

Besides core data management services, *knowledge-based Grids,* built on top of computational and data Grid environments, are needed to offer higher-level services for data analysis, inference, and discovery in scientific and business areas (Moore, 2001). In Berman (2001), Cannataro, Talia, and Trunfio (2001), and Johnston (2002), it is claimed that the creation of knowledge Grids is the enabling condition for developing high-performance knowledge discovery processes and meeting the challenges posed by the increasing demand for power and abstractness coming from complex problem-solving environments.

Several distributed knowledge discovery systems exploiting the Grid infrastructure have been designed and implemented. In this section we discuss some of the most significant ones. The systems discussed here provide different approaches in supporting knowledge discovery on Grids. We discuss them starting from general frameworks, such as the TeraGrid infrastructure, then outlining data-intensive oriented systems, such as InfoGrid and DataCutter, and finally, describing distributed KDD systems such as Discovery Net, and some significant Data Mining testbed experiences.

The *TeraGrid* infrastructure (Catlett, 2002) currently connects 11 sites in the United States to create an integrated computational resource. The resulting Grid environment provides peta-scale data storage and computational capacity. Furthermore, the TeraGrid network connections have been designed in such a way that all resources appear as a single physical site. The connections have also been optimized in order to support peak requirements rather than an average load, as is natural in Grid environments. The TeraGrid adopts Grid software technologies and, from this point of view, appears as a "virtual system" in which each resource describes its own capabilities and behavior through *service specifications*. The most challenging task on TeraGrid is the synthesis of knowledge from very large scientific datasets. This includes, among the others, applications in molecular biosciences, physics, chemistry, and astronomical sciences.

InfoGrid is a service-based data integration middleware engine designed to operate on Grids. Its main objective is to provide information access and querying services to knowledge discovery applications (Giannadakis et al., 2003). The information integration approach of InfoGrid is not based on the classical idea of providing a "universal" query system: instead of abstracting everything for users, it gives a personalized view of the resources for each particular application domain. The assumption here is that users have enough knowledge and expertise to manage with the absence of "transparency." In InfoGrid the main entity is the *Wrapper*; wrappers are distributed on a Grid and each node publishes a directory of the wrappers it owns. A wrapper can wrap information sources and programs or can be built by composing other wrappers (*Composite Wrapper*). In summary, InfoGrid puts the emphasis on delivering metadata describing resources and providing an extensible framework for composing queries.

DataCutter is another middleware infrastructure that aims to provide specific services for the support of multidimensional range querying, data aggregation, and user-defined filtering over large scientific datasets in shared distributed environments (Beynon et al., 2001). In the DataCutter framework, an application is decomposed into a set of processes, called *filters*, which are able to perform a rich set of queries and data transformation operations. Filters can execute anywhere but are intended to run on a machine close (in terms of connectivity) to the storage server. DataCutter supports efficient indexing. In order to avoid the construction of a huge single index that would be very costly to use and keep updated, the system adopts a multilevel hierarchical indexing scheme, specifically targeted at the multidimensional data model adopted.

Differently from the two environments discussed above, the *Datacentric Grid* is a system directed at knowledge discovery on Grids designed for mainly dealing with immovable data (Skillicorn and Talia, 2002). The system consists of four kinds of entities. The nodes at which computations happen are called *Data/Compute Servers* (DCSs). Besides a compute engine and a data repository, each DCS comprises a *metadata tree*, which is a structure for maintaining relationships among raw datasets and models extracted from them. The *Grid Support Nodes* (GSNs) maintain information about the whole Grid. Each GSN contains a directory of DCSs with static and dynamic information about them (e.g., properties and usage), and an execution plan cache containing recent plans along with their achieved performance. The *User Support Nodes* (USNs) carry out execution planning and maintain results. USNs are basically proxies for user interface nodes (called *User Access Points*, UAPs). This is because user requests and their results can be small in size, so in principle

UAPs could be simple devices not always online, and USNs could interact with the Datacentric Grid when users are not connected.

An agent-based data mining framework, called *ADaM* (*Algorithm Development and Mining*), has been developed at the University of Alabama.* Initially, this framework was adopted for processing large datasets for geophysical phenomena. Later, it was ported to the National Aeronautics and Space Administration's (NASA) *Information Power Grid* (IPG) environment for the mining of satellite data (Hinke and Novonty, 2000). ADaM is composed of a moderately rich set of interoperable operation modules, including *data readers* and *writers* for a variety of formats, *preprocessing modules* (e.g., for data subsetting), and *analysis modules* providing data mining algorithms.

The InfoGrid system mentioned earlier has been designed as an application-specific layer for constructing and publishing knowledge discovery services. In particular, it is intended to be used in the *Discovery Net* system (Curcin et al., 2002). Discovery Net is a system developed at the Imperial College whose main goal is to provide a software infrastructure to effectively support scientific knowledge discovery processes from high-throughput informatics. The building blocks in Discovery Net are the *Knowledge Discovery Services* (KDSs) distinguished in *Computation Services* and *Data Services*. The former typically include algorithms (e.g., data preparation and data mining), while the latter define relational tables (as queries) and other data sources. KDSs are used to compose moderately complex data-pipelined processes. The composition may be carried out by means of a graphical user interface (GUI) that provides access to a library of services. Each composed process can be deployed and published as a new process.

The *DataMiningGrid* (Stankovski et al., 2008) is an environment suitable for executing data analysis and knowledge discovery tasks in a wide range of different application sectors, including the automotive, biological and medical, environmental, and information and communications technology (ICT) sectors. Based on open-source Grid middleware, it provides functionality for tasks such as

---

* ADaM, Data Mining and Image Processing Toolkits, http://datamining.itsc.uah.edu/adam (accessed June 2012).

data manipulation, resource brokering, and application searching according to different data mining tasks and methodologies, and supporting different types of functionality for parameter sweeps. In summary, it is a Grid software with all the generic functionality of its component middleware, but with additional features that ease the development and execution of complex data mining tasks.

The *GridMiner* project (Brezany, Janciak, and Min Tjoa, 2008) aims to cover the main aspects of knowledge discovery on Grids. GridMiner is based on the OGSA model and embraces an open architecture in which a set of services are defined for handling data distribution and heterogeneity, supporting different types of analysis strategies, as well as tools and algorithms, and providing for online analytical processing (OLAP) support.

GATES (Grid-Based AdapTive Execution on Streams) is an OGSA-based system that provides support for processing data streams in a Grid environment (Agrawal, 2003). GATES aims to support the distributed analysis of data streams arising from distributed sources (e.g., data from large-scale experiments/simulations), providing automatic resource discovery, and an interface for enabling self-adaptation to meet real-time constraints.

The *Knowledge Grid* (Cannataro and Talia, 2003) exploits the OGSA model to execute distributed knowledge discovery workflows on a Grid environment. Workflows play a fundamental role in the Knowledge Grid at different levels of abstraction. A client application submits a distributed data mining application to the Knowledge Grid by describing it through an XML workflow formalism named *conceptual model*. The conceptual model describes data and algorithms to be used, without specifying information about their location or implementation. The Knowledge Grid creates an *execution plan* for the workflow on the basis of the conceptual model and executes each node of the workflow as a service by using the resources effectively available. The Knowledge Grid will be described in detail in Chapter 5.

Another OGSA-based system for distributed knowledge discovery on the Grid is Weka4WS (Talia, Trunfio, and Verta, 2008). Differently from the Knowledge Grid, which has been implemented from scratch, Weka4WS is an extension of the well-known Weka toolkit.

In the original Weka system, data mining tasks can be executed only locally. Weka4WS extends Weka by allowing remote execution of the algorithms through OGSA-compliant Web services. In Weka4WS, the data mining algorithms for classification, clustering, and association rules can be executed by exploiting remote (computational and data) resources. The Weka4WS framework will be described further in Chapter 6.

We conclude by outlining some interesting data mining testbeds developed at the National Center for Data Mining (NCDM) at the University of Illinois at Chicago (UIC)*:

Terra Wide Data Mining Testbed (TWDM) is an infrastructure for the remote analysis, distributed mining, and real-time exploration of scientific, engineering, business, and other complex data. It consists of a number of geographically distributed nodes linked by optical networks. A central idea in TWDM is to keep generated predictive models up to date with respect to newly available data in order to achieve better predictions (as this is an important aspect in many "critical" domains, such as infectious disease tracking).

Terabyte Challenge Testbed is an open, distributed testbed for DataSpace tools, services, and protocols. It involves a number of organizations, including the UIC, the University of Pennsylvania, the University of California at Davis, and the Imperial College. Each site provides a number of local clusters of workstations that are connected to form wide area *meta-clusters*. So far, meta-clusters have been used by applications in high-energy physics, computational chemistry, nonlinear simulation, bioinformatics, medical imaging, network traffic analysis, digital libraries of video data, and so forth.

---

* http://www.ncdm.uic.edu (accessed June 2012).

CHAPTER 3

Designing Services for Distributed Knowledge Discovery

T HIS CHAPTER EXPLAINS how the different steps of the knowledge discovery in databases (KDD) process can be designed and implemented as services. Moreover, a methodology for composing services to design distributed KDD applications is described. As the approach allows for worldwide knowledge discovery, a way to build a hierarchy of services in large-scale scenarios is discussed.

3.1 A SERVICE-ORIENTED LAYERED APPROACH FOR DISTRIBUTED KDD

Computer science applications are becoming more and more network centric, ubiquitous, knowledge intensive, and computing demanding. This trend will soon result in an ecosystem of pervasive applications and services that professionals and end users can exploit everywhere. As mentioned in the previous chapter, collections of information technology (IT) services and applications, such as Web services, Grid services, and Cloud computing services, are becoming widely available, opening the way for accessing computing services as public utilities, such as water, gas, and electricity.

Key technologies for implementing that perspective include semantic Web and ontologies, ubiquitous computing, peer-to-peer systems, ambient intelligence platforms, data mining and knowledge discovery algorithms and tools, Web 2.0 facilities, mashup tools, and decentralized programming languages. In fact, it is necessary to develop solutions that integrate some or many of those technologies to provide future knowledge-intensive software utilities. The Grid and Cloud paradigms can represent key components of the future Internet, a cyber-infrastructure for efficiently supporting that scenario.

In the area of Grid computing, a proposed approach in accordance with the trend outlined above is the service-oriented knowledge utilities (SOKU) model (NGG3 Expert Group Report, 2005) that envisions the integrated use of a set of technologies that are considered as a solution to information, knowledge, and communication needs of many knowledge-based scientific, industrial, and business applications. The SOKU approach stems from the necessity of providing knowledge and processing capabilities to everyone, thus supporting the advent of a competitive knowledge-based economy. Although the SOKU model is not yet implemented, Grids are increasingly equipped with data management tools, semantic technologies, complex workflows, data mining features, and other Web intelligence approaches. Similar efforts are currently devoted to developing knowledgeable and intelligent Clouds. These technologies can facilitate the process of having Grids and Clouds as strategic components for supporting pervasive knowledge-intensive applications and utilities.

Grids were originally designed for dealing with problems involving large amounts of data and compute-intensive applications. However, Grids have expanded their horizons as they are used to run business applications supporting consumers and end users (Cannataro and Talia, 2004). To face these new challenges, Grid environments must support adaptive knowledge management and data analysis applications by offering resources, services, and decentralized data access mechanisms. In particular, according to the service-oriented architecture (SOA) model, data mining tasks and knowledge discovery processes can be delivered as services in Grid-based or Cloud-based infrastructures.

Through a service-based approach we can define integrated services for supporting distributed KDD applications. Those services can address all the aspects that must be considered in data mining and in knowledge discovery processes such as data selection and transport, data analysis, knowledge models representation, and visualization. Such an approach

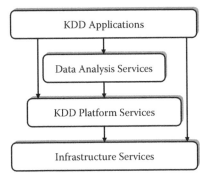

FIGURE 3.1 A service-oriented layered approach for distributed knowledge discovery.

can be described through the layered architecture shown in Figure 3.1, which represents the relationship among *infrastructure services, KDD platform services, data analysis services,* and *KDD applications*:

- *Infrastructure services*: These are core functionalities provided by Grid or Cloud environments. Such functionalities include *security services* (mechanisms for authentication, authorization, cryptography, etc.), *data management services* (data access, file transfer, replica management, etc.), *execution management services* (resource allocation, process creation, etc.), and *information services* (resource representation, discovery, and monitoring).

- *KDD platform services*: These are services specifically designed to support the implementation of knowledge discovery services and applications. They include *resource management services*, which provide mechanisms to describe, publish, and retrieve information about data sources, data mining algorithms, and computing resources; and *execution management services* that allow users to design and execute distributed KDD applications. Examples of KDD platform services are those provided by the Knowledge Grid environment (Cannataro and Talia, 2003), which will be presented and discussed in Chapter 5.

- *Data analysis services*: These are ad hoc services that exploit the KDD platform services to provide high-level data analysis functionalities. A data analysis service can expose either a single step of the KDD process (e.g., data filtering) or data mining task (e.g., data classification), or a more complex knowledge discovery process (e.g., ensemble learning, meta-learning, parallel classification, etc.). Discussion of how data analysis services can be composed to build complex KDD applications is presented in Section 3.2.

- *KDD applications*: These are knowledge discovery applications built upon the functionalities provided by the underlying infrastructure environments, the KDD platform, or higher-level data analysis services. Section 3.3 presents some examples showing how KDD applications can be built following this approach.

This layered approach is flexible in allowing KDD applications to be built upon data analysis services and upon KDD platform services. Moreover, those services can be composed together with the services provided by a Grid or Cloud framework to develop KDD applications that fully exploit the underlying infrastructure features.

Distributed KDD includes a variety of application scenarios relying on many different patterns, techniques, and approaches. Their implementation may involve multiple independent or correlated tasks, as well as the sharing of distributed data and knowledge models (Kargupta and Chan, 2000). Some examples, introduced in Chapter 1, are ensemble learning, meta-learning, and collective data mining.

When such kinds of applications are to be deployed on a distributed environment like the Grid, in order to fully exploit the advantages of the environment, many issues must be faced, including heterogeneity, resource allocation, and execution coordination. A KDD platform should provide services and functionalities that permit coping with most of the above-mentioned issues and the designing of complex applications using high-level abstractions such as service composition facilities (i.e., workflows).

In some cases, however, end users are not familiar with distributed KDD patterns and complex workflow composition. Therefore, higher-level data analysis services, exposing common distributed KDD patterns as single services, might be useful to broaden the accessibility of the KDD platform features.

In the next section, we discuss how KDD applications can be designed as a collection of data analysis services, which include either single steps of the KDD process, single data mining tasks, or more complex knowledge discovery processes involving many independent or correlated tasks.

3.2 HOW KDD APPLICATIONS CAN BE DESIGNED AS A COLLECTION OF DATA ANALYSIS SERVICES

By exploiting service-oriented technologies, such as Web services and the Web Services Resource Framework (WSRF) standards, it is possible to define data analysis services addressing all the aspects that must be considered in the KDD process, from data preprocessing to data mining, knowledge model representation, and visualization.

According to the layered approach introduced above, we can provide data analysis services that can be used at different levels from single operations on data to distributed data mining patterns and complete KDD processes running as service-based workflows on a geographically distributed collection of machines (Talia and Trunfio, 2010a).

This can be done by designing three levels of data analysis services (see Figure 3.2) corresponding to

1. *Single KDD steps and single data mining (DM) tasks*, which include services associated with single operations of the knowledge discovery process (e.g., selection, preprocessing, transformation), and single data mining tasks (e.g., classification, clustering, association rules discovery).

2. *Distributed data mining patterns*, which implement, as services, parallel and distributed data patterns such as ensemble learning, meta-learning, and parallel classification.

3. *Knowledge discovery processes*, which include the previous steps, tasks, and patterns composed in a multistage workflow of services that can be executed on distributed and parallel computing infrastructures.

It is worth noticing that services provided at one level can be used to implement services in other levels; thus, referring to Figure 3.2, single KDD steps services and single DM tasks services can be used to implement distributed data mining patterns, and all three classes of services can be exploited to develop complete knowledge discovery processes. This incremental approach avoids the reimplementation of already available operations, tasks, or patterns and provides a collection of services that can be viewed as a "distributed mining engine."

This collection of data analysis services constitutes an *open service framework* for Grid- and Cloud-oriented knowledge discovery. By exploiting this open framework for service-oriented data mining in Grids,

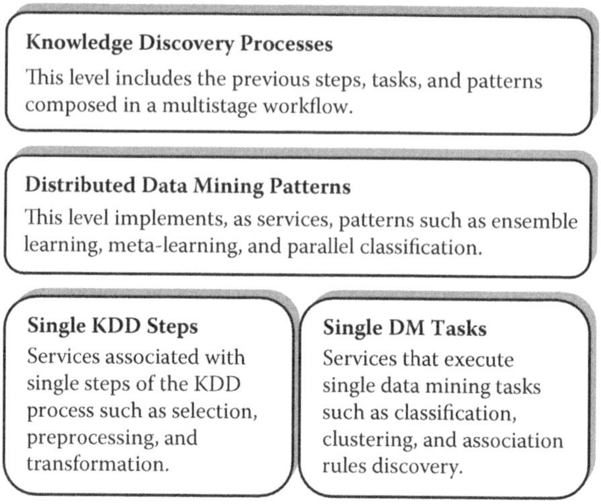

Knowledge Discovery Processes
This level includes the previous steps, tasks, and patterns composed in a multistage workflow.

Distributed Data Mining Patterns
This level implements, as services, patterns such as ensemble learning, meta-learning, and parallel classification.

Single KDD Steps
Services associated with single steps of the KDD process such as selection, preprocessing, and transformation.

Single DM Tasks
Services that execute single data mining tasks such as classification, clustering, and association rules discovery.

FIGURE 3.2 Three levels of data analysis services to facilitate the implementation of distributed knowledge discovery in databases (KDD) applications.

Clouds, and dynamic distributed infrastructures, it is possible to develop data mining services that are accessible every time and everywhere. This solution can support

- Service-based distributed data mining applications

- Data mining services for virtual organizations

- Distributed data analysis services on demand

Therefore, we have a sort of knowledge discovery ecosystem composed of a large number of decentralized data analysis services that will help users to face the availability of massive amounts of data both in business and science.

This approach also facilitates data privacy preservation and prevents disclosure of data beyond the original sources, because it is based on the idea of keeping data at the owner site and not requiring moving data to different locations for its analysis. This data-centric approach is mainly based on moving computation to data rather than moving data to computation. Finally, basic infrastructure services for handling security, trustworthiness, monitoring, and scheduling distributed tasks can be used to provide efficient implementation of high-performance distributed data analysis.

3.3 KDD SERVICE-ORIENTED APPLICATIONS

In the following we present two high-level example scenarios to discuss how distributed KDD applications can be implemented as a collection of data analysis services, according to the layered approach introduced in the previous section. Moreover, we discuss the low-level invocation mechanisms that can be used to enable interactions between long-running KDD services, by exploiting the WSRF family of standards introduced in the previous chapter.

3.3.1 Example Scenarios

As example scenarios, let us consider the case of distributed KDD applications based on two well-known distributed data mining (DDM) patterns, *ensemble learning* and *meta-learning*, already introduced in Section 1.4. These are two significant examples of how DDM services can be implemented through the composition of services associated with single KDD steps and single DM tasks. In fact, from a client-application perspective, ensemble learning and meta-learning patterns should be available as single data analysis services providing distributed learning capabilities through the invocation of lower-level services whose location, interface, and implementation are transparent to the client.

As a first example, we describe a service-oriented ensemble learning scenario. We recall that ensemble learning aims at improving the accuracy of classification by aggregating predictions of multiple classifiers (see Section 1.4.1). We assume that the input dataset is located on a networked storage medium that can be accessed through a storage service. The whole ensemble learning application may be executed as follows (see Figure 3.3):

1. The client invokes an ensemble learning service (ELS), by specifying the identifier of the input dataset to be processed.

2. The ELS starts the ensemble learning process by invoking the storage service, which makes available for processing the input dataset specified by the client.

3. The ELS invokes a partitioning service, which splits the input dataset into a training set and a test set.

4. The ELS invokes n classification services, which take the training set in input and run in parallel to build n independent classification models.

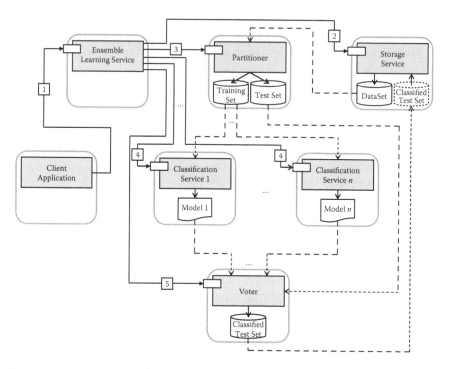

FIGURE 3.3 An ensemble learning application implemented using a service-oriented approach.

5. The ELS invokes a voting service, which performs an ensemble classification by assigning to each instance of the test set the class predicted by the majority of the n classification models generated by the classification services. The classified test set is finally stored through the storage service.

As mentioned above, the second example is a service-oriented implementation of a meta-learning process. In a data classification scenario, meta-learning can be performed by learning from the predictions of a set of base classifiers on a common validation set (see Section 1.4.2). Similar to ensemble learning, meta-learning is a multistep process that can be described as follows (see Figure 3.4):

1. A meta-learning service (MLS) is invoked by the client, by specifying the identifiers of n training sets and one validation set that must be used for the computation.

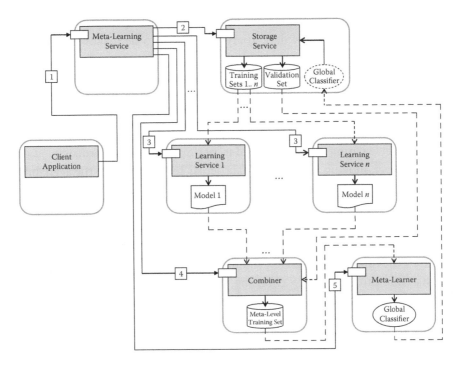

FIGURE 3.4   A service-oriented meta-learning application.

2. The MLS invokes the storage service, which extracts the datasets
   specified by the client from a storage medium.

3. A set of *n* learning services are invoked by the MLS; each learning
   service takes a training set in input and produces an independent
   classification model, or base classifier.

4. The MLS invokes a combining service, which produces a meta-level
   training set by combining the predictions of the base classifiers on
   the provided validation set.

5. The MLS invokes a meta-learner that trains a global classifier from
   the meta-level training set generated in the previous step.

A key role for practical and efficient implementation of these examples is played by the availability of *infrastructure services* provided by
a Grid or a Cloud environment, or more specialized *KDD platform
services* provided by higher-level knowledge discovery frameworks.

As discussed in Section 3.1, both classes of services provide important functionalities that allow data analysis services to be discovered, composed, and executed.

For instance, a Grid information service, like the Globus Toolkit's Monitoring and Discovery Service (Schopf et al., 2005), may be used by a client application to discover the availability of an ELS or MLS, while both ELS and MLS may have to invoke the same information service to find the lower-level learning services needed to perform the overall process.

Moreover, moving and sharing data across services require the availability of an efficient file transfer service, like GridFTP (Allcock et al., 2005), or the use of a distributed file system. In both cases, in the spirit of the approach of moving computation to data rather than moving data to computation, KDD platform services may suggest which services, among the ones discovered by the information service, should be selected to minimize file transfer costs. This can be done by choosing, for example, those services that are physically close to the storage medium where data are originally stored.

3.3.2 Invocation Mechanisms

The example scenarios discussed above demonstrated how data analysis services associated with DDM patterns can be provided through the composition of lower-level data analysis services. A key aspect to enable a practical implementation of such scenarios is the availability of effective invocation mechanisms that support the interactions between long-running services. In fact, data analysis algorithms often take a very long time to complete, and so they cannot be exposed as ordinary Web service operations based on synchronous request and response patterns when they require a long time to produce the knowledge model.

The WSRF, introduced in Section 2.2.2.2, provides a set of basic mechanisms to cope with this aspect. The possibility of defining a state associated with a service is the most important difference between WSRF-compliant and ordinary Web services. This is a key feature in designing distributed KDD applications, because WS-Resources provide a way to represent, advertise, and access properties related to long-running processes. In addition, the WS-Notification specification defines a publish-subscribe notification model for Web services, which is exploited to notify interested clients and services about changes that occur to the status of a resource. The combination of stateful resources and the notification pattern can be exploited to build distributed, long-running applications in which the status of the computation is managed across multiple nodes.

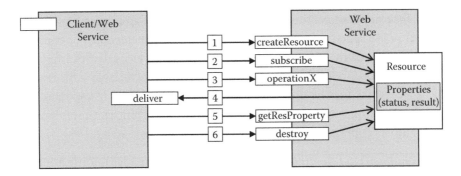

FIGURE 3.5   Interaction between a client and a service using the Web Services Resource Framework (WSRF) with notifications.

To show how stateful resources and the notification model can be exploited in a KDD context, let us assume the scenario in Figure 3.5, in which a client application (or a service acting as a client) wants to invoke a generic *operationX* provided by the service on the right. According to the examples discussed in the previous section, this operation may perform either a simple task (e.g., classification of a single dataset) or a more complex task (e.g., meta-learning on distributed data). Besides the DM-specific *operationX*, the service provides some WSRF-specific operations—*createResource, subscribe, getResourceProperty,* and *destroy*—that are used to manage the computation as detailed in the following.

Using WSRF, the whole invocation process may be divided into six steps (see Figure 3.5):

1. *Resource creation*: The client invokes the *createResource* operation to create a new *resource*. This resource will be used to maintain the state of the computation throughout all the subsequent invocations of the service until its destruction. The state is stored as one or more resource *properties*, which can include specific information such as the current task status (e.g., running, complete, etc.) and the task result (e.g., the inferred model, a pointer to a classified dataset, etc.).

2. *Notification subscription*: The client invokes the *subscribe* operation in order to be notified about changes that will occur to the resource property of interest (e.g., the one containing the task status). Whenever this property value changes, the client will be notified of it. This mechanism will allow the client to be promptly informed about task completion.

3. *Task submission*: The client invokes *operationX* to ask for the execution of a specific data mining task. To perform the invoked operation, the service may in turn invoke some other services. For example, if *operationX* requires a dataset that is not locally available, a file transfer service may be invoked to obtain a local copy of it. As soon as the task completes its execution, the result of the computation is stored into a property of the resource created in the first step.

4. *Notification reception*: As soon as the result is ready, a notification is sent to the client by invoking its implicit *deliver* operation.

5. *Results retrieving*: The client invokes the *getResourceProperty* operation to retrieve the value of the property containing (a pointer to) the computation result.

6. *Resource destruction*: The client finally invokes the *destroy* operation, which eliminates the resource created on the first step.

There are cases in which the notification mechanism cannot be used, for example, due to the presence of a firewall that blocks notifications delivery on the client side. In these cases, a standard polling approach may be used, with the client application that periodically asks the service to know whether the task is still running or has already completed its execution. Using the polling approach, the whole invocation process may be divided into four steps, as shown in Figure 3.6:

1. *Resource creation*: The *createResource* operation is invoked to create a new resource that will be used to maintain task status and result.

2. *Task submission*: The *operationX* is invoked to submit the execution of a specific data mining task. As soon as the task completes its execution, the result is stored into a resource property.

3. *Polling/results retrieving*: The client periodically invokes the *getResourceProperty* operation to retrieve the current status of the computation. When the status indicates task completion, this operation is invoked again to retrieve the property value containing the result.

4. *Resource destruction*: The *destroy* operation is invoked to eliminate the resource created at the beginning of the computation.

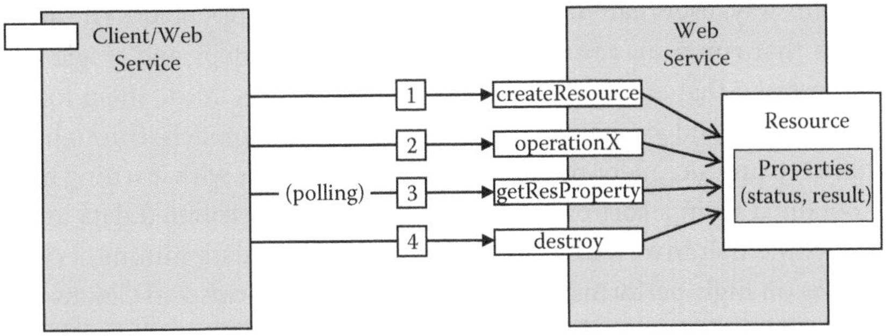

FIGURE 3.6   Interaction between a client and a service using the Web Services Resource Framework (WSRF) without notifications.

Even if polling is simple and can be used in all network scenarios, it may produce either significant delays in result delivery if the polling period is too long or excessive traffic if it is too short. A trade-off solution is using the polling pattern as the default result delivery scheme; when and if it is found that notifications work, may client and service switch to a notification scheme. This hybrid approach for result delivery is adopted by the Weka4WS framework (Talia, Trunfio, and Verta, 2008), which will be described in Chapter 6.

3.4 HIERARCHY OF SERVICES FOR WORLDWIDE KDD

Web services provide a standard way of implementing interoperability among different software applications running on a variety of Web platforms and hardware/software environments connected through the Internet. A very large variety of applications running on remote machines connected over the Internet have been implemented by using Web services. As we discussed in this chapter, this approach can also be exploited in the area of data mining to implement distributed knowledge discovery applications through the composition of services that run tasks corresponding to the different steps of a KDD process and exploit distributed infrastructures such as Web servers, Grids, and Clouds.

This approach supports both single-user distributed KDD processes and multi-user and multi-institution KDD applications that integrate Web services provided worldwide by different players. In fact, these KDD applications can exploit data analysis services deployed by remote third parties and made accessible through the Internet.

In this way, users can develop and deploy service-oriented KDD applications that run a massive number of single KDD steps and single data mining tasks that access data available on Web sites, mine them locally concurrently, and exchange the local knowledge models through the Internet to produce meta-models and to provide a user with learning models obtained from a sort of worldwide large-scale distributed data analysis machine that involves hundreds or thousands of data mining services running on high-performance infrastructures like Grids and Clouds.

Whereas this approach is used today in large-scale traditional Web applications, Web-based information systems, and Web search engines, no work has been done to conceive such a massive data analysis engine based on the composition of services described in Section 3.2.

The development of worldwide data analysis engines by composing single data analysis services and distributed data mining patterns requires the availability of data analysis tasks exposed as Web services and techniques to orchestrate a large number of services that are the building blocks of a distributed data analysis engine. If this is done, new achievements can be reached in the area of computational intelligence because many machine learning algorithms and techniques can be put together to extract knowledge from the everyday increasing data repositories available on the Web. Also, a new generation of knowledge extraction engines can be designed to organize and manage the huge amount of information stored on the Internet.

CHAPTER 4

Workflows of Services for Data Analysis

T HE MAIN GOAL OF THIS CHAPTER is to describe how knowledge discovery in databases (KDD) applications can be programmed as workflows of services running in a distributed environment. We first introduce the basic workflow concepts, focusing on the main patterns used to compose workflow activities. Then, we present some workflow management systems that are relevant to scientific application domains, including data analysis. Finally, we present some examples of distributed KDD workflows and discuss how data analysis applications can be expressed as workflows using graphical representations and coordination languages for their execution.

4.1 BASIC WORKFLOW CONCEPTS

Workflows have emerged as an effective paradigm to address the complexity of scientific and business applications. They provide a declarative way of specifying the high-level logic of an application while hiding the low-level details that are not fundamental for application design.

According to the Workflow Management Coalition, a *workflow* is "the automation of a business process, in whole or part, during which documents, information or tasks are passed from one participant to another for action, according to a set of procedural rules" (1999, 8).

The term *business process* (or, simply, *process*) indicates a set of tasks linked together with the goal of creating a product or providing a service.

Hence, each *task* (or *activity*) represents a piece of work that forms one logical step of the overall process.

The definition, creation, and execution of workflows are supported by a *workflow management system* (WMS). A key function of a WMS during the workflow execution (or *enactment*) is coordinating the operations of the individual activities that constitute the workflow.

4.1.1 Workflow Patterns

Workflow tasks can be composed together following a number of different patterns, whose variety helps designers addressing the needs of a wide range of application scenarios. A comprehensive collection of workflow patterns, focusing on the description of control flow dependencies among tasks, was described in van der Aalst, ter Hofstede, Kiepuszewski, and Barros (2003). In the remainder of this section we focus on a subset of such workflow patterns—sequence, AND/XOR splits, AND/XOR joins, and iteration—that are supported by most workflow languages and that serve as building blocks for more complex workflow patterns.

The simplest workflow pattern is the *sequence*, also called *sequential routing* or *serial routing*. A sequence is a segment of a workflow in which two or more tasks are executed in sequence under a single thread of execution. Hence, a task in a sequence is enabled only after the completion of its preceding task. Figure 4.1 shows an example of a sequence that imposes the sequential execution of three tasks A, B, and C.

A second basic pattern is the *AND-Split*, also called *parallel routing*, *parallel split*, or *fork*. An AND-Split is a point in the workflow where a single thread of control splits into two or more threads of controls that are executed in parallel. This allows multiple tasks in a workflow to be executed concurrently or in any order. Figure 4.2 shows an example of workflow where three tasks, B, C, and D, are executed in parallel after completion of task A thanks to the use of an AND-Split.

An AND-Split is often used in association with an *AND-Join*, which is also called *synchronization point, rendezvous,* or simply *join*. An AND-Join is a point in the workflow where two or more parallel tasks converge into one single thread of control. In other terms, the AND-Join synchronizes

FIGURE 4.1   Sequence.

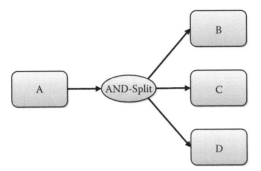

FIGURE 4.2   AND-Split.

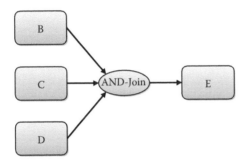

FIGURE 4.3   AND-Join.

multiple tasks before activating the next task. An example is shown in Figure 4.3, where task E is executed only after tasks B, C, and D have completed their execution.

Another pattern is the *XOR-Split*, also called *conditional routing*, *switch*, or *branch*. It is a point in the workflow where a single thread of control makes a decision upon which branch must be taken among two or more alternative branches. An example is shown in Figure 4.4, where, after completion of task A, one task among B, C, and D will be executed, based on which condition holds among *Case 1, Case 2*, and *Case 3*.

A variation of XOR-Split is the *OR-Split*, or *multichoice*, which is a point in the workflow where the thread of control chooses one *or more* branches to follow. This differs from the XOR-Split, where exactly one branch is chosen to take after the decision is made.

A *XOR-Join*, or *asynchronous join*, is often used in association with a XOR-Split. It is a point in the workflow where two or more alternative branches converge, without synchronization, to a single common task. Synchronization is not required because only one of the incoming branches

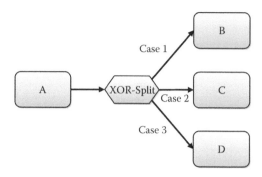

FIGURE 4.4   XOR-Split.

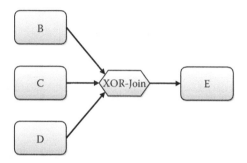

FIGURE 4.5   XOR-Join.

has been followed before the join point. Figure 4.5 shows an example of XOR-Join. Assuming that tasks B, C, and D are placed after a XOR-Split, only one of them will be actually executed. Therefore, as soon as one task among B, C, or D has completed, the execution of task E is started.

If an OR-Split is used in place of a XOR-Split, the XOR-Join can be replaced by a *multimerge*, which is a point in the workflow where two or more branches converge without synchronization. In case multiple branches get activated simultaneously, due to the use of an OR-Split, the task following the multimerge will be started once for every incoming active branch.

An alternative to multimerge is the *discriminator*, or *partial join*. With this pattern, if *m* parallel tasks converge into a single task, it will be activated once *n* tasks out of *m* have completed, whereas the completion of all the remaining tasks will be ignored.

Finally, the *iteration* pattern, also known as *workflow loop* or *cycle*, allows the repetitive execution of one or more workflow tasks until a certain condition is reached. The iteration can be easily implemented using a

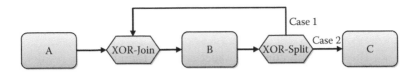

FIGURE 4.6   Example of iteration implemented using a XOR-Join and a XOR-Split.

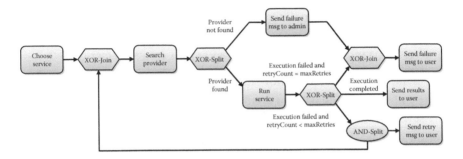

FIGURE 4.7   Example of workflow for service discovery and execution.

XOR-Join and a XOR-Split, as shown in Figure 4.6. In this example, task B is executed repeatedly as long as *Case 1* is true. As soon as the condition indicated as *Case 2* is met, the cycle is stopped and the workflow execution proceeds with task C.

To show how several patterns may be used together to build a complete workflow, Figure 4.7 presents an example of a workflow describing a simple process related to service discovery and execution in a distributed scenario.

The workflow describes, in particular, the following process:

- A user selects a service that he or she wants to invoke from a list of services that are known to be available in a distributed system.

- The request is passed to a discovery service that searches for a provider able to run that service.

- If the discovery service cannot find any provider, a failure message is sent first to the system administrator (to inform him or her that the service is likely to be no longer available in the system), and then to the user (to inform him or her that the service cannot be executed).

- Otherwise, the service is executed by one of the providers found by the discovery service. After execution, three alternative cases are possible:

- The execution fails and the maximum number of retries has been reached; in this case, a failure message is sent to the user.

- The execution completes successfully; therefore, the results produced by the service are delivered to the user.

- The execution fails but the maximum number of retries has not been reached; in this last case, a retry message is sent to the user and an additional iteration is performed.

4.2 SCIENTIFIC WORKFLOW MANAGEMENT SYSTEMS

As mentioned above, workflow management systems (WMSs) are software suites providing tools to define, create, and execute workflows. There are several WMSs on the market, most of them targeted to a specific application domain (e.g., scientific workflows). In the following we focus on some significant WMSs developed by the research community, with the goal of identifying the most important features and solutions that have been proposed for workflow management in the scientific domain.

4.2.1 Taverna

Taverna (Hull et al., 2006) is a Java-based open-source WMS developed at the University of Manchester. The primary goal of Taverna is supporting the Life Sciences community (biology, chemistry, and medicine) in the design and execution of scientific workflows and support *in silico experimentation*, where research is performed through computer simulations with models closely reflecting the real world. Even though most Taverna applications are in the bioinformatics domain, it can be applied to a wide range of fields because it can invoke any Web service by simply providing the URL of its Web Service Description Language (WSDL) document. In addition to Web services, Taverna can invoke local Java services (Beanshell scripts), local Java API (API Consumer), R scripts on an R server (Rshell scripts), and import data from a Cvs or Excel spreadsheet.

The Taverna suite includes four tools:

- *Taverna Engine*, used for enacting workflows.

- *Taverna Workbench*, a client application that enables users to graphically create, edit, and run workflows on a desktop computer.

- *Taverna Server*, enables users to set up a dedicated server for executing workflows remotely.

- *Command Line Tool*, for a quick execution of workflows from a command prompt.

These tools bring together a wide range of features that make it easier to find, design, execute, and share complex workflows. Such features include pipelining and streaming of data, implicit iteration of service calls, conditional calling of services, customizable looping over a service, failover and retry of service calling, parallel execution and configurable number of concurrent threads, and managing previous runs and workflow results. In addition, Taverna provides service discovery facilities and integrated support for browsing curated service catalogs, such as the BioCatalogues.

Finally, even though Taverna was originally designed for accessing Web services and local processes, it can also be used for accessing high-performance computing infrastructures through the use of the TavernaPBS plugin, developed at the University of Virginia, which allows a user to define workflows running on a cluster that uses a portable batch system (PBS).

4.2.2 Triana

Triana (Taylor, Shields, Wang, and Rana, 2004) is a Java-based problem-solving environment, developed at the Cardiff University, which combines a visual interface with data analysis tools. It can connect heterogeneous tools (e.g., Web services, Java units, JXTA services) in one workflow. Triana uses its own custom workflow language, although it can use other external workflow language representations such as Business Process Execution Language (BPEL), which are available through pluggable language readers and writers. Triana comes with a wide variety of built-in tools for signal-analysis, image-manipulation, desktop publishing, and so forth, and has the ability for users to easily integrate their own tools.

The Triana framework is based on a modularized architecture in which the graphical user interface (GUI) connects to a Triana engine, called *Triana Controlling Service* (TCS), either locally or remotely. A client may log into a TCS, remotely compose and run an application, and then visualize the result locally. The application can also be run in batch mode; in this case, a client may periodically log back in to check the status of the application.

The GUI includes a collection of toolboxes containing a set of Triana components, and a workspace where the user can graphically define the required application behavior as a workflow of components. Each component includes information about input and output data type, and the system uses this information to perform design-time-type checking on requested connections in order to ensure data compatibility between components. Several workflow patterns are supported, including loops and branches. Moreover, the workflow components are late bound to the services they represent, thus ensuring a highly dynamic behavior.

Beyond the conventional operational usage, in which applications are graphically composed from collections of interacting units, other usages in Triana include using the generalized writing and reading interfaces for integrating third-party services and workflow representations within the GUI.

4.2.3 Pegasus

The Pegasus system (Deelman et al., 2004), developed at the University of Southern California, includes a set of technologies to execute workflow-based applications in a number of different environments, including desktops, clusters, Grids, and Clouds. Pegasus has been used in several scientific areas including bioinformatics, astronomy, earthquake science, gravitational wave physics, and ocean science.

The Pegasus workflow management system can manage the execution of an application formalized as a workflow by mapping it onto available resources and executing the workflow tasks in order of their dependencies. All the input data and computational resources necessary for workflow execution are automatically located by the system. Pegasus also includes a sophisticated error recovery system that tries to recover from failures by retrying tasks or the entire workflow, by remapping portions of the workflow, by providing workflow-level checkpointing, and by using alternative data sources, when possible. Finally, in order for a workflow to be reproduced, the system records provenance information including the locations of data used and produced, and which software was used with which parameters.

The Pegasus system includes three components:

- The *Mapper* builds an executable workflow based on an abstract workflow provided by the user or generated by the workflow composition system. To this end, this component finds the appropriate software, data, and computational resources required for workflow execution. The Mapper can also restructure the workflow in order to optimize performance, and adds transformations for data management and to generate provenance information.

- The *Execution Engine* executes in appropriate order the tasks defined by the workflow. This component relies on the compute, storage, and network resources defined in the executable workflow to perform the necessary activities.

- The *Task Manager* is in charge of managing single workflow tasks, by supervising their execution on local or remote resources.

4.2.4 Kepler

Kepler (Altintas et al., 2004) is a Java-based open-source software framework providing a GUI and a runtime engine that can execute workflows either from within the graphical interface or from a command line. It is developed and maintained by a team consisting of several key institutions at the University of California, and has been used to design and execute various workflows in biology, ecology, geology, chemistry, and astrophysics.

Kepler is based on the concept of *directors* that dictate the models of execution used within a workflow. Single workflow steps are implemented as reusable *actors* that can represent data sources, sinks, data transformers, analytical steps, or arbitrary computational steps. Each actor can have one or more *input* and *output* ports, through which streams of data tokens flow, and may have *parameters* to define specific behavior. Once defined, Kepler workflows can be exchanged using an eXtensible Markup Language (XML)–based formalism.

By default, Kepler actors run as local Java threads. However, the system also allows spawning distributed execution threads through Web and Grid services. Moreover, Kepler supports foreign language interfaces via the Java Native Interface (JNI), which gives the user flexibility to reuse existing analysis components and to target appropriate computational tools.

The Web and Grid service actors allow users to utilize distributed computational resources in a single workflow. The *WebService* actor provides the user with an interface to easily plug in and execute any WSDL-defined Web service. Kepler also includes a *Web service harvester* for plugging in a whole set of services found on a Web page or in a Universal Description, Discovery, and Integration (UDDI) repository. A suite of *data transformation* actors (XSLT, XQuery, Perl, etc.) allows for semantically compatible but syntactically incompatible Web services to be linked together.

In addition to standard Web services, Kepler includes specialized actors for executing jobs on a Grid. They include actors for certificate-based authentication (*ProxyInit*), submitting jobs to a Grid (*GlobusJob*), as well as Grid-based data transfer (*GridFTP*). Finally, Kepler includes actors for database access and querying: *DBConnect,* which emits a database connection token, to be used by any downstream *DBQuery* actor that needs it.

4.2.5 Askalon

Askalon (Fahringer et al., 2005) is an application development and runtime environment, developed at the University of Innsbruck, which allows the execution of distributed workflow applications in service-oriented Grids. Its service-oriented architecture (SOA)–based runtime environment uses Globus Toolkit as Grid middleware.

Workflow applications in Askalon are described at a high level of abstraction using a custom XML-based language called Abstract Grid Workflow Language (AGWL). AGWL allows users to concentrate on modeling scientific applications without dealing with the complexity of the Grid middleware or any specific implementation technology such as Web and Grid services, Java classes, or software components. Activities in AGWL can be connected using a rich set of control constructs, including sequences, conditional branches, loops, and parallel sections.

The Askalon architecture includes a wide set of services:

- A *resource broker* provides for negotiation and reservation of resources as required to execute a Grid application.

- *Resource monitoring* supports the monitoring of Grid resources by integrating existing Grid resource monitoring tools with new techniques (e.g., rule-based monitoring).

- An *information service* serves as a general-purpose service for the discovery, organization, and maintenance of resource- and application-specific data.

- A *workflow executor* supports dynamic deployment, coordinated activation, and fault-tolerant completion of activities onto remote Grid nodes.

- A *meta-scheduler* performs a mapping of individual or multiple workflow applications onto the Grid.

- *Performance prediction* is a service for the estimation of the execution time of atomic activities and data transfers, as well as of Grid resource availability.

- *Performance analysis* is a service that unifies the performance monitoring, instrumentation, and analysis for Grid applications, and supports the interpretation of performance bottlenecks.

4.3 WORKFLOWS FOR DISTRIBUTED KDD

A key advantage of workflows is the possibility to offload much of the processing to remote components, which makes it feasible to execute compute-demanding workflows from a desktop computer. Moreover, the use of workflows enables the automation of highly repetitive processing stages, which can be modeled as patterns and therefore reused in different applications. These two aspects are very important in compute-intensive data analysis processes that involve decentralization of both data and algorithms, thus making workflows an ideal paradigm to model distributed KDD applications.

To demonstrate the suitability of workflows as a design paradigm for distributed data analysis, in Section 4.3.1 we present some examples of distributed KDD applications expressed as workflows, using the graphical pattern notation introduced in Section 4.1.1. Then, we discuss how a distributed KDD application can be modeled using standard workflow representations languages such as Unified Modeling Language (UML) activity diagrams and BPEL.

4.3.1 Distributed KDD Workflow Examples

As a first example, we consider again the ensemble learning scenario introduced in Section 1.4.1. In particular, Figure 4.8 shows a simple ensemble learning workflow in which the classification is obtained by aggregating the predictions generated on three computing nodes (*node 1*, *node 2*, and *node 3*), by applying three different classification programs. We also assume that the original dataset is stored on a *node 0*; the voting activity is performed on a *node 4*; and the classified test set is stored on *node 0* at the end of the process.

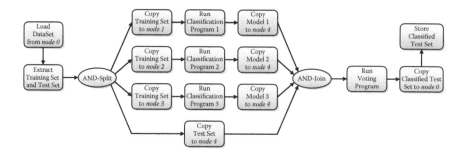

FIGURE 4.8 Example of ensemble learning workflow.

The workflow starts with a sequence of two activities: the first activity loads the input dataset from a storage service located on *node 0*; the second activity splits the input dataset into a training set and a test set.

Then, the workflow proceeds with the parallel execution of four branches mandated by the AND-Split operator. The first parallel branch includes three activities that are performed in sequence: copying the training set to *node 1*; running the first classification program on that training set; and copying the resulting classification model to *node 4*. The second and third parallel branches are identical to the first one, with the difference that the training set is copied to *node 2* (respectively, *node 3*), where it is analyzed using a different classification program and thus generating another classification model. The fourth parallel branch includes one task only: copying the test set to *node 4*.

The four parallel branches synchronize on the AND-Join operator. Therefore, as soon as all four parallel branches have completed their execution, the voting program is run on *node 4*. The voting program will assign to each instance of the test set the class predicted by the majority of the classification models generated on *node 1*, *node 2*, and *node 3*. The classified test set is then copied to *node 0*, where it is finally stored.

As a second example, we consider a workflow that models a parameter sweeping application in which a dataset is analyzed by multiple instances of the same data mining algorithm using different parameters. Parameter sweeping is widely employed in data mining applications to explore the effects of using different values of the algorithm parameters on the results of data analysis.

Figure 4.9 shows, in particular, a parameter sweeping workflow in which a dataset is analyzed in parallel by using three instances of the K-means clustering algorithm. Each K-means instance runs on a different computing node (*node 1, node 2,* or *node 3*) and uses a different value for

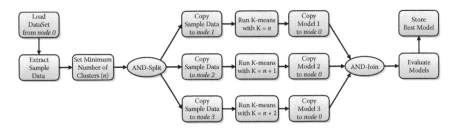

FIGURE 4.9   Example of parameter sweeping workflow.

the parameter K, which represents the number of clusters in which data should be grouped. The goal is to build three different clustering models and to identify the best one among such models. As for the previous workflow, we assume that the original dataset and the final model are both stored on a *node 0*.

The workflow starts with a sequence of three activities: the first activity loads the input dataset from *node 0*; the second activity extracts the sample of data to be analyzed; and the third activity let the user choose the minimum number of clusters, *n*, that must be identified in the sample data.

The workflow proceeds with the concurrent execution of three branches. Each branch includes three sequential activities: copying the sample data to *node 1*, *node 2*, or *node 3*; running K-means with a given value for the parameter K; and copying the resulting clustering model to *node 0*. Parameter sweeping is obtained by setting K = *n* on *node 1*, K = *n* + 1 on *node 2*, and K = *n* + 2 on *node 3*. As soon as the three parallel branches have completed their execution, the three clustering models are evaluated on *node 0* on the basis of their accuracy in order to identify and finally store the best one.

4.3.2 Distributed KDD Workflow Representations

Several graphical representations can be used for modeling workflows, such as UML activity diagrams (Dumas and ter Hofstede, 2001), Event-driven Process Chain (EPC) (van der Aalst, 1999), Petri nets (Murata, 1989), Business Process Model and Notation (BPMN),[*] and directed acyclic graphs (DAGs). The visual representation of a workflow can then be translated into a coordination language for its execution, such as BPEL, or an XML-based ad hoc language.

---

[*] Object Management Group. Business Process Model and Notation, http://www.bpmn.org (accessed June 2012).

As an example, we discuss here how a service-oriented parallel/distributed data mining application can be graphically represented using a UML activity diagram, and how such a representation can be translated into a BPEL document. The UML activity diagram represents the high-level flow of service invocations that constitute the application logic, while BPEL expresses how the various services are actually coordinated and invoked.

The application considered for this purpose is a simple parallel data mining process on distributed data. Three partitions of a dataset, each one located on a different node, contain collections of data to be analyzed by applying a two-phase mining process. Each partition is first processed through the same clustering algorithm that assigns each record to a class. This information is included in the original dataset partition as an additional attribute. The resulting dataset is then processed for classification against the classes obtained on the previous step. The three models are then collected to a node where the best one is chosen by means of a voting operation.

Figure 4.10 shows the UML activity diagram for this example. Note that the *start* activity is followed by a *fork* operator, which specifies that the subsequent activity sequences are to be performed in parallel (AND-Split pattern). The fork operator is then used in combination with a *join* operator, which specifies that the execution flow can proceed only when all of the incoming branches have completed (AND-Join pattern).

After the start, the process runs in parallel on three computing nodes, as mandated by the *fork* operator. On each parallel branch, three sequential activities are specified. The first activity specifies a service invocation requesting the clustering analysis on a dataset partition. The second

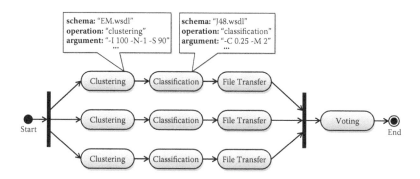

FIGURE 4.10 Unified Modeling Language (UML) activity diagram for a simple parallel/distributed data mining workflow.

activity specifies the classification task to be performed on the dataset resulting from the previous step. Finally, the file transfer activity specifies a data transfer task that copies a classification model to the site where the voting process has to be executed. As specified by the *join*, the voting activity is executed only when all the classification models have been computed and then transferred to the same site.

Many application details, such as the data mining algorithms to be used and their parameters, are hidden behind the visual notation but must be specified by the workflow designer as activity properties. For example, Figure 4.10 shows some properties for two of the data mining tasks (EM for clustering, J48 for classification), which include a reference to the WSDL description of the corresponding service, the specific operation to be invoked, and the operation parameters.

The BPEL notation is explicitly targeted to service invocations and thus is richer than the UML notation. It includes several constructs for interacting with services at different levels, as well as other BPEL processes, and manipulating and accessing service input messages and responses (both of them through explicit variable manipulation operators and XPath expressions).

One important feature about service invocation is the availability of patterns reflecting the typical invocation mechanisms supported by Web services (*one-way, request/response, notification, solicit/response*—see Section 2.2.1.3), which are particularly useful and adaptable to the case of Web Services Resource Framework (WSRF)–compliant Web services.

Figure 4.11 shows the structure of the BPEL document corresponding to the UML activity diagram shown in Figure 4.10. The overall workflow is defined within a *process* element. The *partnerLinks* section defines the services involved in the application. The variables used as input and output in service invocations, as well as for other purposes (e.g., faults and internal variables), are declared within the *variables* section. The *sequence* section specifies the main structure of the application, including a set of parallel activities, and the final voting. The parallel activities are specified using the *flow* operator, which in turn includes three sequences. Each *sequence* is composed of a set of invocations that perform clustering, classification, and transfer tasks. The following is an excerpt of the BPEL document reporting the main invocation steps (*resource creation, notification subscription, task submission, notification reception*—see Section 3.3.2) of one WSRF-compliant Web service exposing the EM clustering algorithm:

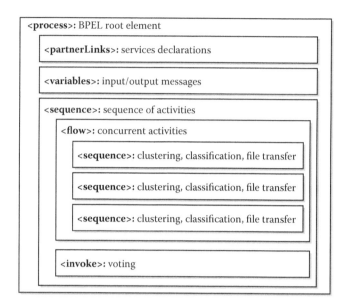

FIGURE 4.11 Structure of the Business Process Execution Language (BPEL) document generated from the Unified Modeling Language (UML) diagram in Figure 4.10.

```
<invoke name="EM service creation"
    partnerLink="EMFactoryService"
    portType="EMFactoryPortType"
    operation="createResource"
    inputvariable="EMCreationRequest"
    outputvariable="EMCreationResponse">
</invoke>
...
<invoke name="EM service subscription"
    partnerLink="EMServiceInstance"
    portType="EMServicePortType"
    operation="subscribe"
    inputvariable="SubscribeInputMessage"
    outputvariable="SubscribeOutputMessage">
</invoke>
...
<invoke name="EM service clustering"
    partnerLink="EMServiceInstance"
    portType="EMServicePortType"
    operation="clustering"
    inputvariable="ClusteringInputMessage">
</invoke>
...
<receive name="EM service notification"
    partnerLink="EMServiceInstance"
    portType="tns:ProcessPortType"
    operation="deliver"
    variable="NotificationMessage">
</receive>
```

The translation of the workflow represented as a UML activity diagram into a BPEL document can be done through a direct mapping between each UML operator and the corresponding BPEL notation, as detailed in Congiusta, Talia, and Trunfio (2008). The overall translation process can be divided into three steps: (1) building the header of the BPEL process, (2) building the body, and (3) composing the previous two sections to form the final BPEL document. The second step is the core of the algorithm and is performed by a *BuildActivity* function that receives a generic UML activity and returns the corresponding BPEL activity.

By exploiting object-oriented inheritance mechanisms, all the BPEL operators can be defined in a hierarchy, whose root is represented by a BPELActivity class. Therefore, every BPEL operator (for example, BPELSequence, BPELFlow, etc.) is defined as a subclass of BPELActivity. The same applies to UML operators, which inherit from a base UMLActivity class.

BuildActivity operates in a recursive way, by detecting the actual type of the UML operator and applying a set of mapping rules. When the function receives a composite UML activity, it proceeds recursively to build the corresponding BPEL activity. For example, the UML *fork* operator is translated into a BPEL *flow* block, whose elements are obtained by invoking BuildActivity on the *fork* subactivities.

On the other hand, when the function receives a basic UML activity, it can build the corresponding BPEL *invoke* operator through a direct translation. This requires that the properties specified for a UML activity be translated into the appropriate attributes in the BPEL *invoke* domain. For example, the input and output parameters specified in the UML activity are used to build input and output variables needed by BPEL in the service invocation.

CHAPTER 5

Services and Grids

The Knowledge Grid

THIS CHAPTER PRESENTS THE KNOWLEDGE GRID SYSTEM, an open-source environment for designing and running distributed knowledge discovery in databases (KDD) applications as workflows. The chapter describes the Knowledge Grid's service-oriented architecture, outlines how software and data resources are modeled in the system, describes the DIS3GNO tool for composing KDD workflows in the Knowledge Grid, and explains how such workflows are executed on a Grid platform.

5.1 THE KNOWLEDGE GRID ARCHITECTURE

The Knowledge Grid framework makes use of basic Grid services to build more specific services to support distributed KDD on the Grid (Cannataro and Talia, 2003). Such services allow users to implement knowledge discovery applications that involve data, software, and computational resources available from distributed Grid sites. To this end, the Knowledge Grid defines mechanisms and higher-level services for publishing and searching information about resources, representing KDD applications, and managing their distributed execution.

As mentioned in Section 2.3.1, workflows play a fundamental role in the Knowledge Grid at different levels of abstraction. A client application submits a distributed KDD application to the Knowledge Grid by describing it through a workflow formalism named *conceptual model*. The conceptual model describes data and algorithms to be used, possibly without specifying information about their location or implementation.

The Knowledge Grid creates an *execution plan* for the workflow on the basis of the conceptual model and executes each task of the workflow

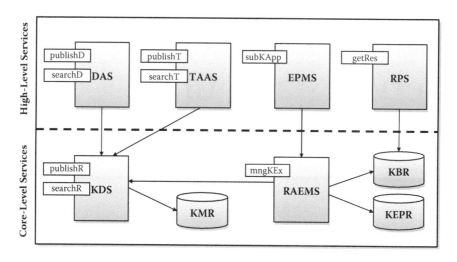

FIGURE 5.1 The Knowledge Grid architecture: layers, services, and operations.

using the software and hardware resources available in the system. To this end, the Knowledge Grid follows a two-step approach: it initially builds an *abstract execution plan* that in a second step is resolved into a *concrete execution plan*. The abstract execution plan may not contain specific information about the involved resources (e.g., the actual Grid node where a certain data mining algorithm will run), while in the concrete execution plan all the resources must be actually specified by finding a mapping between requested resources and the ones available in the system.

5.1.1 Core Services and High-Level Services

The Knowledge Grid services are organized in two hierarchical layers: the *high-level layer* and the *core-level layer*, as shown in Figure 5.1. The design idea is that client applications interact directly with high-level services that, in order to satisfy client requests, invoke suitable operations exported by the core-level services. In turn, core-level services perform their operations by invoking basic services provided by the underlying Grid environment, as well as by interacting with other core-level services.

The high-level layer includes the following services:

- The *Data Access Service* (DAS) provides operations for publishing and searching data to be mined.

- The *Tools and Algorithms Access Service* (TAAS) is responsible for publishing and searching tools and algorithms for data extraction, preprocessing, and data mining.

- The *Execution Plan Management Service* (EPMS) receives a conceptual model of the KDD application, translates it into an abstract execution plan, and passes it to the core-level RAEMS service (described below) for its execution.

- The *Results Presentation Service* (*RPS*) allows the results of previous KDD computations to be retrieved.

The core-level layer includes two services:

- The *Knowledge Directory Service* (KDS) is responsible for managing metadata about the Knowledge Grid resources (data, tools, and algorithms). It provides operations to publish and search resource metadata, which are stored in a *Knowledge Metadata Repository* (KMR).

- The *Resource Allocation and Execution Management Service* (RAEMS) starts from an abstract execution plan, received by the high-level EPMS service, generates a concrete execution plan, and manages its execution. The generated execution plans are stored in a *Knowledge Execution Plan Repository* (KEPR), while the computation results are stored in a *Knowledge Base Repository* (KBR).

The design approach conceived the Knowledge Grid architecture as a set of core-level and high-level services that do not impose any constraint on the implementation strategy. However, to fully exploit existing Grid middleware and to simplify and homogenize client-to-service and service-to-service interactions, all the Knowledge Grid services have been implemented as WSRF-compliant Web services (Congiusta, Talia, and Trunfio, 2007). In particular, the current Knowledge Grid is based on the WSRF Java library provided by Globus Toolkit 4 (Foster, 2005).

The invocation mechanism of the Knowledge Grid services follows the publish-subscribe pattern discussed in Section 3.3.2, as described in Figure 3.5. This means that each Knowledge Grid service, besides the specific operations related to the functions it provides, exposes some WSRF-specific operations (*createResource*, *subscribe*, *getResourceProperty*, and *destroy*—see Section 3.3.2) that are required to manage stateful resources, subscribe to notifications, and get the results. The main specific operations provided by the different Knowledge Grid services are listed and described in Table 5.1.

TABLE 5.1   Main Operations Implemented by the Knowledge Grid Services

Service	Operation	Description
DAS	*publishData*	This operation is invoked to publish a newly available dataset. The publication requires a set of information that will be stored as metadata in the local KMR.
	searchData	Data to be used in a KDD application can be searched and located by invoking this operation. The search is performed on the basis of appropriate parameters.
TAAS	*publishTool*	This operation is used to publish metadata about a data mining tool into the local KMR. As a result of the publication, the new software is made available for use in KDD applications.
	searchTool	It is similar to the *searchData* operation except that it is targeted to search data mining tools.
EPMS	*submitKApplication*	This operation receives a conceptual model of the KDD application to be executed. The EPMS generates the corresponding abstract execution plan and submits it to the RAEMS for its execution.
RPS	*getResults*	This operation retrieves the results of a KDD application and presents them to the user.
KDS	*publishResource*	This is the core-level operation for publishing data or tools. It is invoked by the DAS or TAAS services to perform their specific operations.
	searchResource	This is the core-level operation for searching data or tools.
RAEMS	*manageKExecution*	This operation receives an abstract execution plan of the KDD application. The RAEMS generates a concrete execution plan and manages its execution.

5.2 METADATA MANAGEMENT

The effective use of a Grid requires the definition of a model to manage the heterogeneity of the involved resources, which can include computers, data, network facilities, sensors, and software tools generally provided by different organizations (Mastroianni, Talia, and Trunfio, 2004). Heterogeneity in Grids arises mainly from the large variety of resources available within each category. For example, a software package can run only on some particular host machine, whereas data can be extracted from different data management systems such as relational databases, semistructured databases, plain files, and so forth.

The management of such heterogeneous resources requires the use of metadata, whose purpose is to provide information about the features of resources and their effective use. A Grid user needs to know which resources are available, where resources can be found, how resources can be accessed, and when they are available. Metadata can provide answers about involved computing resources such as data repositories (e.g., databases, file systems, Web sites), machines, networks, programs, documents, user agents, and so on. Therefore, metadata represent a key element for the effective discovery and utilization of resources on the Grid.

The Knowledge Grid defines a set of metadata schemas for two main classes of resources: *data mining software*, which includes tools and algorithms for data analysis and manipulation, and *data sources,* such as plain files, databases, semistructured documents, and other structured or unstructured data. Metadata are expressed using eXtensible Markup Language (XML), due to its extendibility, platform independence, and for the ease in mapping XML documents into the data structures of object-oriented programming languages.

5.2.1 Metadata Representation

The Knowledge Grid addresses the problem of giving metadata a suitable schema for properly representing information about the different types of software and data resources that may be involved in a distributed KDD application.

A first step in designing a metadata schema for data mining software is their classification. In the Knowledge Grid, data mining software are categorized on the basis of the following parameters (Chen, Han, and Yu, 1996):

- The kind of input data sources.

- The kind of knowledge that is to be discovered.

- The type of techniques that data mining software tools use in the mining process.

- The driving method of the mining process.

Table 5.2 summarizes a set of values for each classification parameter. This table is mapped on an XML Schema that defines the format and the syntax of the XML file that will be used to describe the features of a generic data mining software. The second column of the table reports the XML elements corresponding to the classification parameters.

TABLE 5.2   Classification of Data Mining Software

Classification Parameter	eXtensible Markup Language (XML) Tag	Possible Values
Type of data sources	`<KindOfData>`	Relational database, transaction database, object-oriented database, deductive database, spatial database, temporal database, multimedia database, heterogeneous database, active database, legacy database, semistructured data, flat file
Type of knowledge to be mined	`<KindOfKnowledge>`	Association rules, clusters, characteristic rules, classification rules, sequence discovery, discriminant rules, evolution analysis, deviation analysis, outlier detection, regression
Type of techniques to be utilized	`<KindOfTechnique>`	Statistics, decision trees, neural networks, genetic algorithms, Apriori, fuzzy logic, singular value decomposition (SVD), Bayesian networks, nearest neighbors
Driving method	`<DrivingMethod>`	Autonomous knowledge miner, data-driven miner, query-driven miner, interactive data miner

As an example, the following is an extract of an XML metadata document describing a command-line instance of the AutoClass data mining software (Cheeseman and Stutz, 1996):

```
<DataMiningSoftware name="AutoClass">
  <Description>
    <KindOfData>flat file</KindOfData>
    <KindOfKnowledge>clusters</KindOfKnowledge>
    <KindOfTecnique>statistics</KindOfTecnique>
    <DrivingMethod>autonomous knowledge miner</DrivingMethod>
  </Description>
  <Usage>
    ...
    <Invocation>/usr/autoclass/autoclass</Invocation>
    <Args>
      <Arg name="search" type="string" mandatory="true">-search</Arg>
      <Arg name="db2 file" mandatory="true"/>
      <Arg name="hd2 file" mandatory="true"/>
      <Arg name="model file" mandatory="true"/>
      <Arg name="s-params file" mandatory="true"/>
      ...
    </Args>
    <Hostname>gridlab1.deis.unical</Hostname>
    <ManualPath>/usr/autoclass/read-me.text</ManualPath>
    ...
  </Usage>
</DataMiningSoftware>
```

The XML document is composed of two parts. The first part is the software *Description*, and the second is the software *Usage*. The *Description* section specifies one or more values, among those reported in Table 5.2, for each classification parameter. The *Usage* section contains information that can be used by a client application to access and use the software package. This section includes the software invocation details and the associated arguments.

As mentioned before, data sources can originate from plain files, relational databases, Web pages, and other structured and semistructured documents. In spite of the wide variety of possible data source types, the Knowledge Grid defines a common structure of data source metadata in order to homogenize the access and search operations on such resources. The common structure of metadata is composed of two parts:

- An *Access* section that includes information for retrieving a data source.

- A *Structure* section that provides information about the logical and physical structure of a data source.

As an example, the following is an extract of the XML metadata document for a flat file that can be used as an input by the AutoClass software:

```
<FlatFile>
  <Access>
    <Location>/usr/share/imports-85c.db2</Location>
    <Size>26756</Size>
    ...
  </Access>
  <Structure>
    <Format>
      <AttributeSeparatorString>,</AttributeSeparatorString>
      <RecordSeparatorString>#</RecordSeparatorString>
      <UnknownTokenString>?</UnknownTokenString>
      ...
    </Format>
    <Attributes>
      <Attribute name="symboling" type="discrete">
        <SubType>nominal</SubType>
        <Parameter>range 7</Parameter>
      </Attribute>
      <Attribute name="normalized-loses" type="real">
        <SubType>scalar</SubType>
        <Parameter>zero_point 0.0</Parameter>
        <Parameter>rel_error 0.01</Parameter>
      </Attribute>
      ...
    </Attributes>
  </Structure>
</FlatFile>
```

The *Access* section includes file system information (e.g., the file *Location* and *Size*). The *Structure* section includes two subsections, *Format* and *Attributes*. The *Format* subsection contains information about the physical structure of the flat file (e.g., the strings that are used to separate the records and the attributes within a record). The *Attributes* subsection contains information about the logical structure (i.e., it lists the table attributes and provides the relative specifications, such as the *name* of the *Attribute*, its *type*, and so forth).

Although the high-level XML metadata format is the same for all kinds of data sources, the content of the *Access* and *Structure* sections may depend on the specific characteristics of a given data source. As an example, for relational databases the *Format* subsection is not needed because the physical formatting is managed by the database management system. Furthermore, new subsections can be introduced; for instance, in the *Access* section, information should be provided for the connection to the database (e.g., the open database connectivity [ODBC] specifications).

5.2.2 Metadata Publication and Search

Metadata publication and search are managed by the core-level KDS service, while the high-level DAS and TAAS services provide to client applications specialized operations for publishing and searching data and tools, respectively, as detailed in Table 5.1. DAS and TAAS possess the same basic structure and perform their main tasks by interacting with a local instance of the KDS, which in turn may invoke one or more other remote KDS instances.

Figure 5.2 describes the interactions occurring when the DAS service is invoked; similar interactions apply also to TAAS invocations. The *publishData* operation is invoked to publish information about a dataset (Step 1). The DAS passes the corresponding metadata document to the local KDS by invoking the *publishResource* operation (Step 2). The KDS, in turn, stores the metadata document in the local KMR (Step 3).

The search for a data source is performed through the *searchData* operation starting from a search string passed by the client (Step 4). This string contains the searching criteria expressed as attribute-value pairs regarding key properties through which data sources are categorized within the system by using the metadata model described above. The DAS submits the request to the local KDS by invoking the corresponding *searchResource* operation (Step 5). As mentioned before, the KDS performs the search by querying both the local KMR and a set of remote KDSs (Step 6). This is a

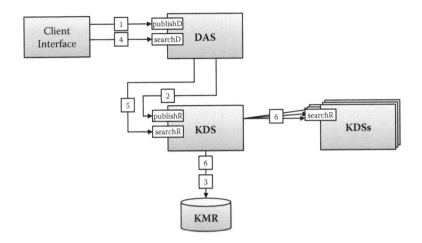

FIGURE 5.2   Interactions between Knowledge Grid services for metadata publication and search.

general rule enforced in all the interactions between a high-level service and the KDS when a search is requested. The local KDS is thus responsible for dispatching the query to remote KDSs and generating the final answer, which is a set of URLs (referred to as KDS URLs in Knowledge Grid's terminology) pointing to the metadata of the data sources satisfying the search criteria.

To reach as many remote KDSs as needed, an unstructured peer-to-peer overlay similar to Gnutella (Ripeanu, Iamnitchi, and Foster, 2002) is built among the Knowledge Grid nodes. The peer-to-peer overlay is constructed by assigning to each node a small set of neighboring nodes. Each node sends a KDS query to its neighbors, which in turn can forward it to their neighbors to ensure wider network coverage. As in Gnutella, this query flooding is controlled in two ways: (1) each time a query is forwarded, an associated time-to-leave (TTL) counter is decremented by one; when the TTL value equals zero, the query forwarding is stopped. (2) If a node receives the same query more than once, as a consequence of looping paths that may be present in the overlay, the query is discarded without further processing.

5.3 WORKFLOW COMPOSITION USING DIS3GNO

Within the Knowledge Grid project, a visual client interface named DIS3GNO has been implemented (Cesario, Lackovic, Talia, and Trunfio, 2011) to allow users to

- Program distributed data mining workflows

- Execute the workflow onto the Knowledge Grid as a collection of services

- Visualize the results of a data mining task

DIS3GNO performs the mapping of the user-defined workflow to the conceptual model and submits it to the Knowledge Grid services, managing the overall computation in a way that is transparent to a user.

In supporting a user to develop applications, DIS3GNO is the interface for two main Knowledge Grid functionalities:

- *Metadata management*: DIS3GNO provides an interface to publish and search metadata about data and tools, through the interaction with the DAS and TAAS services.

- *Design and execution management*: DIS3GNO provides an environment to program and execute distributed KDD applications as service-oriented workflows, through the interaction with the EPMS service.

The DIS3GNO graphical user interface (GUI), shown in Figure 5.3, has been designed to reflect this twofold functionality. In particular, it provides a panel (on the left) devoted to search resource metadata, and a panel (on the right) to compose and execute KDD workflows.

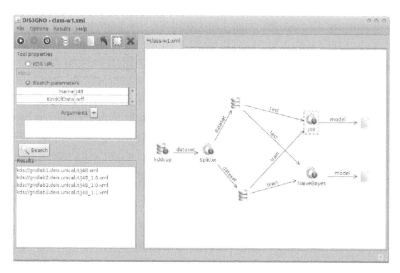

FIGURE 5.3   A screenshot of the DIS3GNO graphical user interface (GUI).

In the top-left corner of the window there is a menu used for opening, saving, and creating new workflows; viewing and modifying some program settings; and viewing the previously computed results stored in the local file system. Under the menu bar, there is a toolbar containing some buttons for the execution control (starting/stopping the execution and resetting the nodes statuses) and others for the workflow editing (creation of nodes representing datasets, tools, or models; creation of edges; selection of multiple nodes; and deletion of nodes or edges).

5.3.1 Workflow Representation

In DIS3GNO, a workflow is represented as a directed acyclic graph whose nodes represent resources and whose edges represent dependencies among the resources. The types of resources that can be present in a KDD workflow (graphically depicted by the icons in Figure 5.4) are as follows:

- *Dataset*, representing a dataset.

- *Tool*, representing a tool to perform any kind of operation that may be applied to a dataset (data mining, filtering, splitting, etc.) or to a model (e.g., voting operations).

- *Model*, representing a knowledge model (e.g., a decision tree, a set of association rules, etc.) (i.e., the result produced by a data mining service).

Data mining services can be either implemented by users or taken from third-party software. A reliable source of data mining algorithms is the Weka toolkit (Witten and Frank, 2000), from which are taken most of the algorithms implemented as services in the DIS3GNO experiments that will be described in Chapter 8. Besides data mining algorithms, Weka provides a wide set of useful filtering tools that can also be used to query remote datasets and to create a local copy of them for subsequent processing.

Each node contains a description of a resource as a set of properties providing information about its features and actual usage. This description may be complete or partial: in other words, it is either possible to

DataSet Node   Tool Node   Model Node

FIGURE 5.4   Node types.

specify a particular resource and its location in the Grid, or just a few of its properties, thus leaving to the system the task of finding a resource that matches the required properties. In the former case we refer to the resource as *concrete*, and in the latter one as *abstract*.

For example, in the case of a data mining tool, one could be interested in any algorithm, located in any Grid node, provided it is a classification algorithm able to handle "arff" files, or could want specifically the algorithm named "NaiveBayes" located in a specified host. Once the workflow will be run, the Knowledge Grid services will find a concrete resource matching the metadata, whether they are completely or partially specified. Clearly, only dataset and tool nodes can be either concrete or abstract, while a model node cannot be abstract as it represents a computation result. The model node has only one property—the location—which if left empty will be implicitly set to the same location of the tool node in input.

When a particular resource property is entered, a label is attached below to the corresponding icon, as shown in the examples in Figure 5.5. The property chosen as the label is the one considered most representative for the resource (i.e., the name for the dataset and tool nodes and the location for the model node).

In order to ease the workflow composition and allow a user to monitor its execution, each resource icon bears a symbol representing the status in which the corresponding resource is at a given time. When the resource status changes, as a consequence of the occurrence of certain events, its status symbol changes accordingly. The resource statuses can be divided into two categories: the composition-time and the runtime statuses.

The composition-time statuses (shown in Table 5.3) are useful during the workflow composition phase. They are as follows:

1. *No information provided*: No parameter has been specified in the resource properties.

2. *Abstract resource*: The resource is defined through constraints about its features, but it is not a known Apriori; the **S** in the icon stands for search, meaning that the resource has to be searched in the Grid.

Iris   K-Means   gridlab6

FIGURE 5.5 Node labels.

TABLE 5.3   Node Composition-Time Statuses

Symbol	Meaning
⚙	No information provided
⚙Ⓢ	Abstract resource
⚙Ⓚ	Concrete resource
Ⓛ	Location set

3. *Concrete resource*: The resource is specifically defined through its KDS URL; the *K* in the icon stands for KDS URL.

4. *Location set*: A location for the model has been specifically set; this status (*L*) is pertinent to model nodes only.

The runtime statuses (shown in Table 5.4), which are useful during the workflow execution phase, are as follows:

1. *Matching resource found*: A concrete resource matching the metadata has been found.

2. *Running*: The resource is being executed/managed.

3. *Resource not found*: The system has not found a resource matching the metadata.

4. *Execution failed*: Some fault has occurred during the management of the corresponding resource.

5. *Task completed successfully*: The corresponding resource has successfully fulfilled its task.

TABLE 5.4   Node Runtime Statuses

Symbol	Meaning
⚙▶	Matching resource found
⚙↻	Running
⚙✕	Resource not found
⚙⚠	Execution failed
⚙✓	Task completed successfully

TABLE 5.5   Node Connections

First Resource	Second Resource	Label	Meaning	Graphical Representation
Dataset	Dataset	transfer	Explicit file transfer	transfer
Dataset	Tool	dataset, train, test	Type of input for a tool node	?
Tool	Dataset	dataset	Dataset produced by a tool	dataset
Tool	Model	model	Model produced by a data mining algorithm	model
Model	Tool	model	Model received by a tool	model
Model	Model	transfer	Explicit transfer of a model	transfer

Each resource may be in one of these runtime statuses only in a specific phase of the workflow execution (i.e., status 1 and 2 only during the execution; status 3 and 4 during or after the execution; status 5 only after the execution).

The nodes may be connected to each other through edges, establishing dependency relationships among them using specific patterns. The workflow patterns currently supported in DIS3GNO are *sequence* (a task is started after the completion of the preceding task), *parallel split* (in any node with multiple outgoing edges, the thread of control is split into multiple ones, thus allowing parallel execution), and *join* (any node with multiple incoming edges is implicitly a point of synchronization) (van der Aalst et al., 2003). All the possible connections are shown in Table 5.5; connections not included in Table 5.5 are not allowed, and the GUI prevents a user from creating them.

When an edge is being created between two nodes, a label is automatically attached to it representing the kind of relationship between the two nodes. In most cases this relationship is strict, but in one case (dataset-tool connection) it requires further input from a user to be specified.

The possible edge labels are

- *Dataset*: Indicates that the input or output of a tool node is a dataset.

- *Train*: Indicates that the input of a tool node has to be considered a training set.

- *Test*: Indicates that the input of a tool node has to be considered a test set.

- *Transfer*: Indicates an explicit transfer of a dataset, or a result of a computation, from one Grid node to another.

- *Model*: Indicates a result of a computation of a data mining algorithm.

5.3.2 Workflow Composition

To outline the main functionalities of DIS3GNO, we briefly describe how it is used to compose and run a distributed KDD workflow. By exploiting the DIS3GNO GUI, a user can compose a workflow by selecting from the toolbar the type of resource to be inserted in the workflow (a dataset, a tool, or a model node), and clicking on the workflow composition panel. Such an operation can be repeated as many times as needed to insert all the required application nodes. Then, he or she has to insert suitable edges by setting, for each one, the specific dependency relationship between the nodes (as described in Section 5.3.1 and summarized in Table 5.5). Typically, most nodes in a workflow represent abstract resources. In other terms, a user initially focuses on the application logic, without focusing on the actual datasets or data mining tools to be used.

Let us suppose that a user wants to compose and execute the ensemble learning application described in Section 1.4.1 and shown in Figure 1.11, in which (1) the input dataset is split, using a partitioner tool, into a training set and a test set; (2) the training set is given in input to n classification algorithms that run in parallel to build n independent classification models from it; and (3) a voter tool performs an ensemble classification by assigning to each instance of the test set the class predicted by the majority of the N models generated at the previous step.

By using DIS3GNO, the ensemble learning application can be designed as follows. First, a user chooses the input dataset (Figure 5.6). To do that, he or she selects from the toolbar the dataset icon and drags it into the workflow composition panel. For associating the icon to a concrete resource, the user specifies the name and format of the desired dataset and sets it into the Search parameters panel. The search is started by pressing the Search button. On completion, the system lists the set of datasets matching the search criteria (left-bottom part of the GUI).

After the selection of one dataset, the URL of such dataset is associated with the dataset icon (Figure 5.7a). This operation changes the dataset icon mark from *S* (resource still to be selected) to *K* (resource identified by a KDS URL). In the same way, the user inserts a tool node that is associated with the KDS URL of the desired partitioner (Figure 5.7b). Then, the user

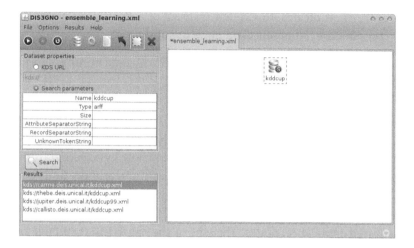

FIGURE 5.6   Insertion of the input dataset icon with specification of its properties and search for matching resources.

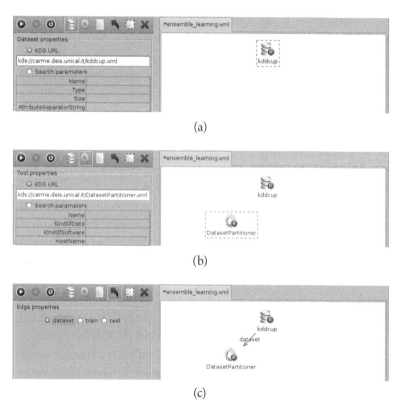

(a)

(b)

(c)

FIGURE 5.7   (a) Selection of the input dataset. (b) Insertion and selection of a partitioner tool. (c) Insertion of a labeled edge between dataset and partitioner.

must specify the relationship between the two nodes. To do that, an edge linking the dataset and the tool icons is created and labeled appropriately (Figure 5.7c).

According to the ensemble learning scenario, two dataset icons representing the output of the partitioner are added to the workflow (Figure 5.8a). Then, the user proceeds by adding the classification algorithms that are required to build the base models. We assume that the user wants to use four classification algorithms (ConjunctiveRule, NaiveBayes, RandomForest, and J48) specified as abstract resources (see Section 5.3.1). For example, Figure 5.8b shows the insertion of the first classification algorithm (ConjunctiveRule) and the specification of its properties (name of software and type of data supported). The algorithm icon is marked with an *S* to remind that the corresponding resource will be searched and made concrete at runtime. Similarly, the other three classification algorithms are added, and an edge between the training set and the four algorithms is created (Figure 5.8c).

Figure 5.9 shows the complete workflow. It includes (1) a model node connected to each classification algorithm, (2) a tool node representing a voter that takes as input the test set and the four base models, and (3) the output dataset obtained as output of the voter tool.

The workflow can be submitted to the EPMS service by pressing the Run button on the toolbar. As a first action, if user credentials are not available or have expired, a Grid Proxy Initialization window is loaded. After that, the workflow execution actually starts and proceeds as detailed in the next section.

5.4 EXECUTION MANAGEMENT

Starting from the data mining workflow designed by a user, the client interface (DIS3GNO) generates an XML representation of the KDD application referred to as the conceptual model. DIS3GNO passes the conceptual model to a given EPMS, which is in charge of transforming it into an abstract execution plan for subsequent processing by the RAEMS. The RAEMS receives the abstract execution plan and creates a concrete execution plan. To accomplish this task, the RAEMS needs to evaluate and resolve a set of resources and services by contacting the KDS and choosing those matching the requirements specified by the abstract execution plan. In case multiple resources match such requirements, the RAEMS adopts a round-robin strategy to select one of them.

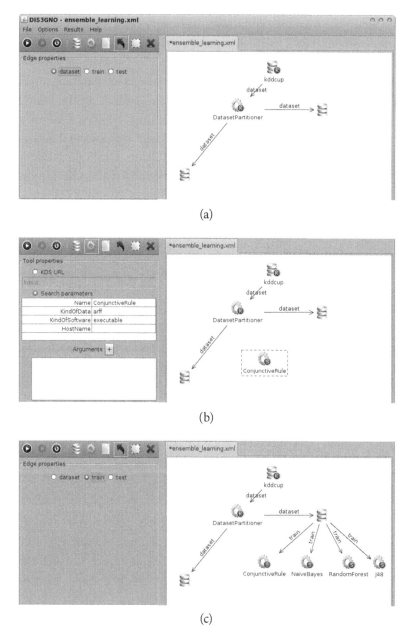

FIGURE 5.8   (a) Insertion of two dataset icons representing the partitioner output. (b) Insertion and specification of an abstract tool resource. (c) Workflow after insertion and specification of all the classification algorithms and associated input edges.

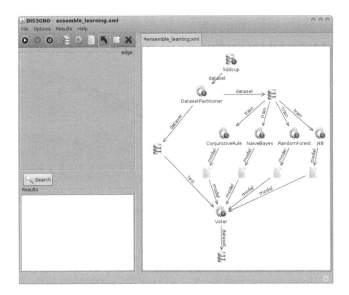

FIGURE 5.9 The complete ensemble learning workflow including the base models, a voter tool, and the output dataset.

As soon as the RAEMS has built the concrete execution plan, it is in charge of invoking the coordinated execution of programs and services associated with the nodes of the concrete execution plan. The status of the computation is notified to the EPMS, which in turn forwards the notifications to the DIS3GNO system for visualization.

Figure 5.10 describes the interactions that occur when an invocation of the EPMS is performed. In particular, the figure outlines the sequence of invocations of other services, and the interchanges with them when a KDD workflow is submitted for allocation and execution. To do this, the EPMS exposes the *submitKApplication* operation, through which it receives a conceptual model of the application to be executed (Step 1).

As an example, in the following we report an extract of the conceptual model generated by DIS3GNO starting from the simple workflow shown in Figure 5.11:

```
<graphml xmlns="http://graphml.graphdrawing.org/xmlns/graphml">
  <graph id="G" edgedefault="directed">
    <node id="n0">
      <data key="type">dataset</data>
      <data key="description">
        <Dataset href="kds://globus1.deis.unical.it/CoverType.arff"/>
      </data>
      <data key="position_X">310</data>
      <data key="position_Y">230</data>
```

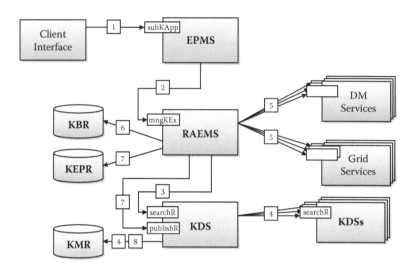

FIGURE 5.10   Interactions between Knowledge Grid services for the execution of a knowledge discovery in databases (KDD) workflow.

FIGURE 5.11   Simple workflow example.

```
    </node>
    <node id="n1">
      <data key="type">algorithm</data>
      <data key="description">
        <DataMiningSoftware name="J48">
          <Description>
            <KindOfData>arff</KindOfData>
          </Description>
        </DataMiningSoftware>
      </data>
      ...
    </node>
    <node id="n2">
      <data key="type">model</data>
      <data key="location">localhost</data>
      ...
    </node>
    <edge id="e0" source="n0" target="n1">
      <data key="type">train</data>
    </edge>
    <edge id="e1" source="n1" target="n2">
      <data key="type">model</data>
    </edge>
  </graph>
</graphml>
```

Basically, the conceptual model is a textual representation of the graph expressed by the visual workflow, which is suitable to be passed to the EPMS for further processing. Conceptual models are expressed using the GraphML formalism that is widely employed to represent graphs in XML. Note that the conceptual model reflects the fact that the original workflow contains a concrete resource (the input dataset CoverType) and an abstract one (the J48 algorithm). In fact, node *n0* in the conceptual model is fully specified by its KDS URL, which refers to the metadata descriptor of the CoverType dataset. On the other hand, node *n1* specifies only the algorithm name (J48) and the kind of data to be processed (arff), thus keeping the resource abstract and giving to the system the task of mapping it to a concrete resource.

As mentioned earlier, the basic role of the EPMS is to transform the conceptual model into an abstract execution plan for subsequent processing by the RAEMS. An abstract execution plan is a more formal representation of the application structure. Generally, it does not contain information on the physical Grid resources and services to be used, but rather constraints about them. In the following, we report an example of an abstract execution plan generated by the EPMS starting from the conceptual model shown above:

```
<AbstractExecutionPlan>
  <Task label="START"/>
  <Task name="sub_job1">
    <Computation>
      <Input href="kds://gridlab1.deis.unical.it/CoverType.xml" id="input1"/>
      <Program id="program1">
        <DataMiningSoftware name="J48">
          <Description>
            <KindOfData>arff</KindOfData>
          </Description>
          <Usage>
            <Args>
              <Arg name="InputTrainingSet">input1</Arg>
              <Arg name="OutputModel">output1</Arg>
            </Args>
          </Usage>
        </DataMiningSoftware>
      </Program>
      <Output id="output1" name="sub_job1_model.model"/>
    </Computation>
  </Task>
  <Task label="END"/>
  <TaskLink from="START" to="sub_job1"/>
  <TaskLink from="sub_job1" to="END"/>
</AbstractExecutionPlan>
```

The RAEMS exports the *manageKExecution* operation, which is invoked by the EPMS and receives an abstract execution plan (Step 2). First, the RAEMS queries the local KDS (through its *searchResource* operation) to find the resources needed to instantiate the abstract execution plan (Step 3). The KDS performs the search by both accessing the local Knowledge Metadata Repository (KMR) and querying a set of remote KDSs, as described in Section 5.2.2 (Step 4).

In the following, we present an example of a concrete execution plan generated by the RAEMS starting from the abstract execution plan shown earlier:

```
<ConcreteExecutionPlan">
  <Task name="START"/>
  <Task name="sub_job1_inputTransfer0">
    <DataTransfer name="sub_job1_inputTransfer0"
      src="gridlab1.deis.unical.it" dest="gridlab2.deis.unical.it">
      <Input href="kds://gridlab1.deis.unical.it/CoverType.xml"/>
    </DataTransfer>
  </Task>
  <Task name="sub_job1">
    <Computation>
      <Input href="kds://gridlab1.deis.unical.it/CoverType.xml">
      </Input>
      <Program href="kds://gridlab2.deis.unical.it/J48.xml">
        <DataMiningSoftware name="J48">
          <HostName>gridlab2.deis.unical.it</HostName>
        </DataMiningSoftware>
      </Program>
      <Output name="sub_job1_model.model"/>
    </Computation>
  </Task>
  <Task name="END"/>
  <TaskLink from="START" to="sub_job1_inputTransfer0"/>
  <TaskLink from="sub_job1_inputTransfer0" to="sub_job1"/>
  <TaskLink from="sub_job1" to="END"/>
</ConcreteExecutionPlan>
```

Note that in the concrete execution plan, the J48 algorithm is instantiated to a concrete resource that is fully specified by its KDS URL. Moreover, compared to the abstract execution plan, the concrete one includes an additional *DataTransfer* operation that consists of copying the input dataset to the node where the selected J48 instance is located. In fact, while the data transfer operation is implicit in the original workflow, it must be explicitly specified in the concrete execution plan.

After the concrete execution plan is obtained, the RAEMS coordinates the actual execution of the overall computation. To this purpose, the RAEMS invokes the appropriate data mining tools and services (DM Services) and the needed Grid Services (e.g., file transfer services) as

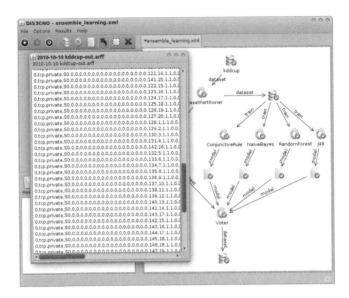

FIGURE 5.12 The final visualization of DIS3GNO after completion of the ensemble learning application.

specified by the concrete execution plan (Step 5). The RAEMS stores the results of the computation into the Knowledge Base Repository (KBR) (Step 6), while the execution plan is stored in the Knowledge Execution Plan Repository (KEPR) (Step 7). To make available the results stored in the KBR, it is necessary to publish results metadata into the KMR. To this end, the RAEMS invokes the *publishResource* operation of the local KDS (Steps 7 and 8).

Figure 5.12 shows the final screenshot of the DIS3GNO interface when the execution of the ensemble learning workflow introduced above is complete and the final classification results have been created and shown in an ad hoc window.

CHAPTER 6

Mining Tasks as Services

The Case of Weka4WS

W EKA4WS EXTENDS THE WELL-KNOWN WEKA DATA MINING TOOL-
KIT to support the execution of distributed knowledge discovery
applications using a service-oriented approach. This chapter introduces
Weka and presents the architecture and features of the Weka4WS frame-
work. The chapter shows, in particular, how the two visual interfaces of
Weka4WS, the Weka4WS Explorer and KnowledgeFlow environments,
can be used for running distributed knowledge discovery in database
(KDD) applications on a Grid infrastructure.

6.1 ENABLING DISTRIBUTED KDD IN AN OPEN-SOURCE TOOLKIT

The Knowledge Grid framework, described in Chapter 5, is an example
of how a complete data analysis system can be implemented as a collection of services providing all the functionalities required to build distributed KDD applications, from data and algorithm management, to
workflow composition and execution. In this chapter we discuss an alternative design approach that has been followed for the development of the
Weka4WS system (Lackovic, Talia, and Trunfio, 2009a; Talia, Trunfio,
and Verta, 2005, 2008). Rather than building a distributed KDD system
from scratch, Weka4WS extends a well-known data analysis framework,
the Weka toolkit (Hall et al., 2009; Witten and Frank, 2000), which was
originally designed to support data mining tasks on a single computer,
making it able to support the execution of distributed knowledge discovery applications on a Grid environment.

Weka4WS stands for *Weka for Web Services*, meaning that all the Weka data mining algorithms can be executed remotely and concurrently through Web service interfaces. Like the Knowledge Grid, Weka4WS adopts the Web Services Resource Framework (WSRF) as core Web service technology and uses Globus Toolkit for basic Grid functionalities, such as authentication, authorization, file transfer, and execution management.

There are three main reasons behind the choice of Weka as a data mining system to extend: it is a well-established software, cross-platform (written in Java), and open source (available under the GNU General Public License). Above all, the fact that Weka is available as open-source software greatly motivated and facilitated the development of Weka4WS, which is in turn provided to the research community under the same licensing terms. The remainder of this section provides an overview of the Weka toolkit and introduces the design goals of the Weka4WS system.

6.1.1 Weka: An Overview

Weka is an open-souce data mining toolkit, developed at the University of Waikato, that provides a large collection of algorithms for data preprocessing, classification, clustering, association rules discovery, and results visualization. These algorithms can be accessed through a common graphical user interface (GUI) that includes four specialized environments:

- *Simple CLI*: This is a command-line interface that can be used for the direct execution of Weka commands in those operating systems that do not provide their own command-line interface.

- *Experimenter*: This is an environment for performing experiments and conducting statistical tests between learning schemes. It can be used to automatically find which methods and parameter values work best for a given problem.

- *Explorer*: It is the main user interface of the Weka toolkit. It provides integrated access to all the Weka algorithms for data preprocessing (loading, filtering, sorting, normalization, etc.), data mining (classification, clustering, association rules discovery), and visualization of results (models inferred by the data mining algorithms).

- *KnowledgeFlow*: It has essentially the same features of the Explorer, with additional support to incremental learning and a drag-and-drop interface that allows composing and executing data mining workflows. By using this interface, a user can select a set of visual components from a tool bar, place them onto a panel, and connect them together to form a "knowledge flow" for processing and analyzing data. Each component represents an individual tool for data preprocessing, data mining, or results visualization.

All the algorithms are written in Java and packed into an extensible *Weka library*. A user can easily include new algorithms in the Weka library, provided that such algorithms respect some interfacing constraints to ensure interoperability with the rest of the system.

6.1.2 Weka4WS: Design Goals

The objective that guided the design of Weka4WS is to allow users to perform distributed data mining on the Grid in an easy and effective way. In particular, the design goals of the system are supporting both

- The execution of one or more independent data mining tasks on remote Grid nodes.

- The execution of multiple data mining tasks, composed in a knowledge discovery workflow, on multiple nodes of a Grid.

Supporting remote execution of single or multiple data mining tasks allows users to exploit the computational power and the distribution of data and algorithms of a Grid to obtain mining results in a shorter time, and to access and mine larger data repositories.

To make as easy as possible the use of the system, Weka4WS builds upon a well-established data mining environment—the Weka toolkit—and extends it with remote execution features. In this way, domain experts can focus on designing their knowledge discovery applications, without worrying about learning complex tools or languages for Grid submission and management. The Weka4WS visual interface allows users to set up their data mining tasks or knowledge discovery workflows as in the original Weka system, with the additional capability of specifying the Grid nodes as to where to execute the data mining algorithms. To enable remote invocation, Weka4WS exposes all the data mining algorithms originally provided by Weka as a WSRF-compliant Web service, which can be easily deployed on the available Grid nodes to set up a distributed execution environment.

Other Grid systems, already mentioned in Section 2.3.1, share with Weka4WS the use of a service-oriented approach to support distributed data mining, including Discovery Net, GridMiner, and the Knowledge Grid. However, the design approach of Weka4WS is different from that of those systems. In fact, Weka4WS aims at extending a widely used framework to minimize the efforts needed by users to learn how to utilize it, while the objective of the systems mentioned above is to implement complete frameworks providing ad hoc services and languages to perform knowledge discovery on a Grid.

Distributed KDD applications can be built in Weka4WS using either the Explorer or the KnowledgeFlow interface: with the former, it is possible to run one or more independent data mining tasks on remote Grid nodes; with the latter, multiple data mining tasks, composed in a knowledge discovery workflow, can be executed on multiple Grid nodes. In both cases, the location where a given data mining task will be executed can be either specified by the user or automatically chosen by the Weka4WS system, as detailed in the remainder of the chapter.

6.2 WEKA4WS ARCHITECTURE

In Weka4WS, nodes are classified into two categories on the basis of the available Weka4WS components: *user nodes*, which are the local computers that provide the Weka4WS client software, and *computing nodes*, computers that receive data mining task submissions from user nodes and manage their execution. Data to be mined can be located on computing nodes, user nodes, or third-party nodes (e.g., shared data repositories). If a dataset to be mined is not present on the computing node where the analysis has to be performed, it is transparently copied to that node using GridFTP (Allcock et al., 2005), a high-performance, secure, and reliable data transfer protocol that is part of Globus Toolkit 4 (GT4). Figure 6.1 shows the software components of user nodes and computing nodes in the Weka4WS framework.

Computing nodes include two components: a *Web service* and the *Weka library*. The Web service exposes all the data mining algorithms provided by the underlying Weka library. Therefore, requests submitted to the Web service are executed by invoking the appropriate algorithms of the Weka library.

User nodes include three components: a *GUI*, the *Weka library*, and a *Client module*. As mentioned before, the GUI includes an extended version of the Weka Explorer and KnowledgeFlow interfaces to enable the distributed execution of single data mining tasks and KDD workflows,

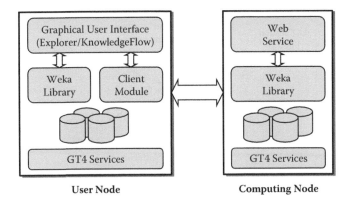

FIGURE 6.1   Software components of user nodes and computing nodes.

respectively. Local tasks are executed by directly invoking the local Weka library, while remote tasks are executed through the client module that operates as an intermediary between the GUI and Web services exposed by the remote computing nodes.

On both user nodes and computing nodes, the appropriate GT4 services must be available. In particular, on user nodes it is sufficient to have present the Globus Java WS Core, a lightweight cross-platform version of GT4 mostly supporting client-side operations, while a complete installation of GT4 is required on each computing node to support full server-side capabilities.

6.2.1 Communication Mechanisms

In the best network scenario, the communication between a client module and a Web service is based on the "push-style" mode of the *notification message* delivery mechanism defined by WS-Notification (see Section 3.3.2), where the client module is the *notification consumer*, and the Web service is the *notification producer*. The client module invokes the service and waits to be notified as soon as the required task has completed its execution.

However, there are certain circumstances in which the push-style mode of the notification message delivery mechanism does not work, for example, when the client resides behind a NAT Router/Firewall, as shown in Figure 6.2, because the messages sent to the client will be blocked unless they are over a connection initiated by the client.

When a client subscribes to notification for a task completion, it also passes to the service a randomly generated port number, to which the client will be listening for receiving the notification: from that moment on, the client will act as a server, waiting for notification to the given port.

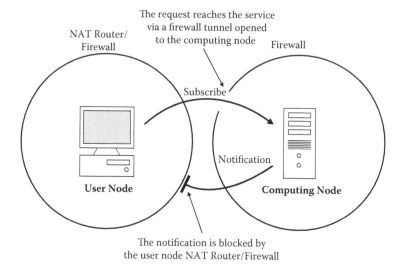

FIGURE 6.2 Notifications are blocked when the client runs on a machine behind a NAT Router/Firewall.

When the results are ready at the computing node, a connection is initiated by the computing node but the attempt to send the notification fails because the user node NAT Router/Firewall blocks it.

In this scenario, the only way for the client module to get results is to work in "pull-style" mode—that is, to continuously try to pull (retrieve), at certain time intervals, the results from the computing node until they will be available. In order to make the system automatically adapt to all possible network scenarios in a way that is transparent to the user, the client module of Weka4WS starts in pull-style mode by default. At the moment of notification subscription, it also asks the service for the immediate delivery of a "dummy" notification whose sole purpose is to check whether the client is able to receive notifications: when and if the client module will receive this dummy notification, it will switch to the push-style mode, otherwise it will persist in the pull-style mode.

6.2.2 Web Service Operations

In each computing node, a Web service answers the user node requests by invoking the appropriate data mining algorithms in the underlying Weka library. The algorithm invocation is performed in an asynchronous way—that is, the client submits the task in a nonblocking mode, and the results are either notified to it as soon as they are ready (push-style mode)

TABLE 6.1   Operations Provided by Each Web Service in the Weka4WS Framework

Operation	Description
classification	Submits the execution of a classification task.
clustering	Submits the execution of a clustering task.
associationRules	Submits the execution of an association rules task.
createResource	Creates a new stateful resource.
subscribe	Subscribes to notifications about resource properties changes.
getResourceProperty	Retrieves the resource property values.
destroy	Explicitly requests destruction of a resource.
getVersion	Returns the version of the service.
notifCheck	Checks whether the client is able to receive notifications.
stopTask	Explicitly requests the termination of a given task.

or they are repeatedly checked for readiness by the client (pull-style mode) depending on the network configuration, as described before.

Table 6.1 lists the operations provided by each Web service in the Weka4WS framework. The first three operations—*classification, clustering*, and *associationRules*—are used to require the execution of a specific data mining task. In particular, the *classification* operation provides access to the complete set of classification algorithms in the Weka library (currently, 71 algorithms). The clustering and association rules operations expose all the clustering and association rules algorithms provided by the Weka library (five and two algorithms, respectively). Note that new algorithms can be added to Weka4WS by simply extending the Weka library, like in the original Weka system (see Section 6.1.1). The four operations in the middle rows (*createResource, subscribe, getResourceProperty, destroy*) are related to the WSRF-specific invocation mechanisms already described in Section 3.3.2. The last three operations (*getVersion, notifCheck, stopTask*) provide some supporting functionalities: the *getVersion* operation is invoked to check whether the client and service versions are compatible; the *notifCheck* operation is invoked just after a *subscribe* operation to check whether the client is able to receive notifications, as described before; finally, *stopTask* requests the termination of a previously submitted data mining task.

The parameters required by the *classification, clustering*, and *associationRules* operations are shown in Table 6.2. The *taskID* field is required solely for the purpose of stopping a task. The *algorithm* field is of a complex type, detailed in Table 6.3, which contains two subfields: *name*, a string identifying the Java class in the Weka library to be invoked (e.g., *weka.*

TABLE 6.2   Input Parameters of the Data Mining Operations

Parameter Type	Field Name	Field Type
classificationParameters	*taskID*	long
	algorithm	algorithmType
	dataset	datasetType
	testset	datasetType
	classIndex	int
	testOptions	testOptionsType
	evalOptions	evalOptionsType
clusteringParameters	*taskID*	long
	algorithm	algorithmType
	dataset	datasetType
	testset	datasetType
	classIndex	int
	testOptions	testOptionsType
	selectedAttributes	array of int
associationRulesParameters	*taskID*	long
	algorithm	algorithmType
	dataset	datasetType

TABLE 6.3   Web Services Input Parameters Field Types

Field Type	Subfield Name	Subfield Type
algorithmType	*name*	string
	parameters	string
datasetType	*fileName*	string
	filPath	string
	dirPath	string
	crc	long
testOptionsType	*testMode*	int
	numFolds	int
	percent	int
evalOptionsType	*costMatrix*	string
	outputModel	boolean
	outputConfusion	boolean
	outputPerClass	boolean
	outputSummary	boolean
	outputEntropy	boolean
	rnd	int

classifiers.trees.J48), and *parameters*, a string containing the sequence of arguments that must be passed to the algorithm (e.g., *–C 0.25 –M2*). The *dataset* and *testset* are two other fields of a complex type that contains four subfields: *fileName*, a string containing the file name of the dataset (or test set); *filePath*, a string containing the full path (file name included) of the dataset; *dirPath*, a string containing the path (file name excluded) of the dataset; and *crc*, which specifies the checksum of the dataset (or test set) file.

The *classIndex* field is an integer designating which attribute of the dataset must be considered as the class attribute when invoking a classification or clustering algorithm. The *testOptions* field is of a complex type containing three subfields: *testMode* is an integer representing the test mode to be applied to the classification or clustering algorithm (1 for cross-validation, 2 for percentage split, 3 to use the training set, and 4 to use a separate test set); *numFolds* and *percent* are two optional fields used when applying a cross-validation or a percentage split test mode, respectively.

The *evalOptions* field is used specifically for the classification algorithms and contains several subfields like *costMatrix* (to evaluate errors with respect to a cost matrix), *outputModel* (to output the model obtained from the full training set), and others shown in Table 6.3. The field *selectedAttributes* is specifically used for the clustering algorithms and contains those data attributes that must be ignored when performing the clustering task.

6.2.3 Client Application

On a user node, the client application starts with the three windows shown in Figure 6.3:

- The GUI Chooser (left side in Figure 6.3) is used to launch Weka's four graphical interfaces.

- The Remote-hosts list-checking Window (top right side in Figure 6.3) is used to give a visual confirmation of the host checking procedure (described below).

- The Grid Proxy Initialization Window (middle right side in Figure 6.3) is automatically loaded at startup only if the user credentials (needed to submit tasks to the remote Grid nodes) are not available or have expired.

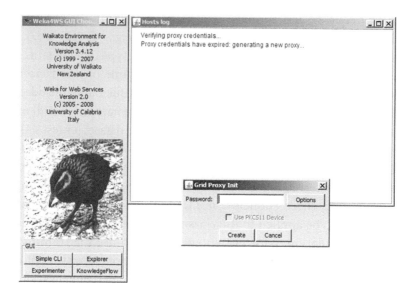

FIGURE 6.3   Weka4WS graphical user interface (GUI) startup.

The remote hosts' addresses are loaded from a configuration file located inside the application directory. The configuration file is read in the background when the application is launched and each remote host is checked to see that:

- The Globus container and GridFTP server are running.

- The Weka4WS Web service is deployed.

- The Weka4WS versions of client and Web service are compatible.

Only those hosts that pass all the checks are made available to the user in the GUI. In order to take into account possible alterations of the Grid network configuration without having to restart the application, the remote hosts' addresses may be reloaded at any time by pressing a given button provided for the purpose.

After remote host checking and credential initialization, a user can choose which interface to launch. In the following two sections, we will describe in more detail the two interfaces—Explorer and KnowledgeFlow—that have been extended in the Weka4WS framework.

6.3 WEKA4WS EXPLORER FOR REMOTE DATA MINING

Explorer, the main GUI in Weka, is a comprehensive tool with six tabbed panes, each one dedicated to a specific Weka facility like data preprocessing

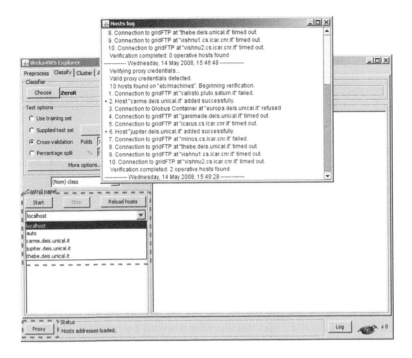

FIGURE 6.4   Weka4WS Explorer: Control Panel and hosts reloading.

(loading from file, URL or database, filtering, saving, etc.), data mining (classification, clustering, association rules discovery), and data visualization.

In Weka4WS, the Explorer component is essentially the same as the Weka one with the exception of the three tabbed panes associated with classification, clustering, and association rules discovery: in those panes the two buttons for starting and stopping the algorithms have been replaced with a Control Panel, and a button named Proxy has been added in the lower-left corner of the window. The modifications are highlighted in Figure 6.4.

The drop-down menu in the Control Panel allows us to choose either the exact Grid location where we want the current algorithm to be executed (where *localhost* will make the algorithm be computed on the local machine) or to let the system automatically choose one by selecting the *auto* entry. The strategy used in the auto mode is round-robin: on each invocation, the host in the list next to the previously used one is chosen.

The Reload hosts button, when pressed, brings up the hosts' list-checking window and starts the hosts' checking procedure, while the Proxy button, when pressed, brings up the Grid Proxy Initialization window, as described in Section 6.2.3.

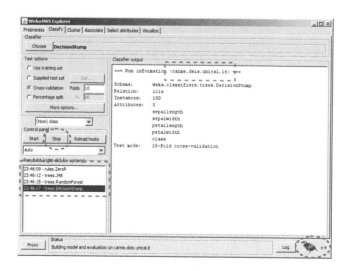

FIGURE 6.5   Weka4WS Explorer: Multiple tasks execution.

Once the Grid node is chosen, be that local or remote, the task may be started by pressing the Start button and stopped by pressing the Stop button. In Weka4WS, unlike in Weka, a task is carried out in a thread of its own, thus allowing multiple tasks to be run in parallel on multiple computing nodes. In Figure 6.5, in the lower-right corner of the window, the number of running tasks is displayed. The list of started tasks is displayed in the Result list pane, just below the Control panel.

The Output panel, at the right side of the window, shows the run information and results (as soon as they are known) of the task currently selected in the Result list; at the top of the Output Panel, as highlighted in Figure 6.5, the host name of the Grid node where the task is being executed is shown.

It is possible to follow the remote computations in their single steps as well as to know their execution times through a Log window, which is shown by pressing the Log button in the lower-right corner of the window, as highlighted in Figure 6.6.

6.4 WEKA4WS KNOWLEDGEFLOW FOR COMPOSING DATA MINING SERVICES

KnowledgeFlow is the Weka interface that allows a user to compose workflows for processing and analyzing data. A workflow can be built by selecting components from a tool bar, placing them on a layout canvas, and

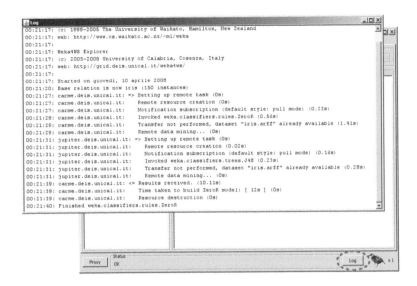

```
Log
00:21:17: (c) 1999-2005 The University of Waikato, Hamilton, New Zealand
00:21:17: web: http://www.cs.waikato.ac.nz/~ml/weka
00:21:17:
00:21:17: Weka4WS Explorer
00:21:17: (c) 2005-2008 University of Calabria, Cosenza, Italy
00:21:17: web: http://grid.deis.unical.it/weka4ws/
00:21:17:
00:21:17: Started on giovedi, 10 aprile 2008
00:21:20: Base relation is now iris (150 instances)
00:21:27: carme.deis.unical.it: => Setting up remote task (0s)
00:21:27: carme.deis.unical.it:    Remote resource creation (0s)
00:21:27: carme.deis.unical.it:    Notification subscription (default style: pull mode) (0.23s)
00:21:28: carme.deis.unical.it:    Invoked weka.classifiers.rules.ZeroR (0.56s)
00:21:29: carme.deis.unical.it:    Transfer not performed, dataset "iris.arff" already available (1.41s)
00:21:29: carme.deis.unical.it:    Remote data mining... (0s)
00:21:31: jupiter.deis.unical.it: => Setting up remote task (0s)
00:21:31: jupiter.deis.unical.it:    Remote resource creation (0.02s)
00:21:31: jupiter.deis.unical.it:    Notification subscription (default style: pull mode) (0.16s)
00:21:31: jupiter.deis.unical.it:    Invoked weka.classifiers.trees.J48 (0.23s)
00:21:31: jupiter.deis.unical.it:    Transfer not performed, dataset "iris.arff" already available (0.28s)
00:21:31: jupiter.deis.unical.it:    Remote data mining... (0s)
00:21:39: carme.deis.unical.it: <= Results received. (10.11s)
00:21:39: carme.deis.unical.it:    Time taken to build ZeroR model: [ 12s ] (0s)
00:21:39: carme.deis.unical.it:    Resource destruction (0s)
00:21:40: Finished weka.classifiers.rules.ZeroR

                                         Status
                           Proxy         OK                                    Log    x 1
```

FIGURE 6.6   Weka4WS Explorer: Detailed log.

connecting them together: each component of the workflow is demanded to carry out a specific step of the knowledge discovery process.

Weka4WS extends the KnowledgeFlow by adding annotations to the KDD workflows. Through annotations, a user can specify how the workflow nodes associated with data mining algorithms must be mapped onto Grid nodes, so as to enable their distributed execution.

Figure 6.7 shows an example of workflow and highlights the main changes introduced in Weka4WS as compared to Weka. The workflow represents a simple KDD process in which a dataset (iris) is analyzed in parallel using one clustering algorithm (Cobweb) and two classification algorithms (Decision Stump and J48). The workflow starts (on the left) with an ArffLoader node, used to load the dataset, which is connected to three nodes: *TrainTestMaker, TestSetMaker,* and *CrossValidationFoldMaker,* which provide the test sets and training sets to the three algorithms, as appropriate. The first two algorithms are connected to a *PerformanceEvaluator* node for the model validation and then to a *TextViewer* node. The third algorithm is directly connected to a *GraphViewer.* The user can see the results of a given data mining task by clicking on the corresponding *TextViewer* or *GraphViewer* node. When the application is started, the three branches of the workflow are executed in parallel on a set of computing nodes.

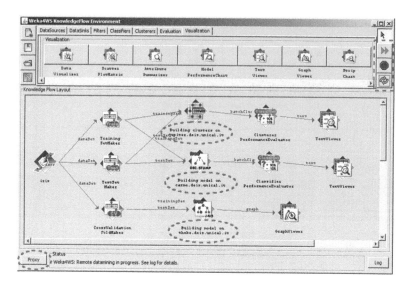

FIGURE 6.7 Weka4WS KnowledgeFlow: Example of workflow and extended features.

Compared to the original Weka KnowledgeFlow, the following changes have been made in Weka4WS (see Figure 6.7): (1) in the upper-right corner three buttons have been added whose purpose is, from top to bottom, to start all the tasks at the same time, to stop all the tasks, and to reload the hosts list, as seen in Section 6.2.3; (2) a button named Proxy has been added in the lower-left corner of the window, which when pressed, just like in the Explorer component, brings up the Grid Proxy Initialization window described earlier; and (3) the labels under each workflow node associated to an algorithm indicate, during the computation, the Grid hosts where that algorithm is being computed.

The choice of the location where to run a certain algorithm is made in the configuration panel of each algorithm, accessible by right-clicking on the given algorithm and choosing Configure, as shown in Figure 6.8: within the highlighted area it can be seen that the part added in Weka4WS, as previously seen from the Control Panel of the Explorer component, consists of a drop-down menu containing the available locations where the selected algorithm can be executed.

Although the algorithms and their performance evaluators are represented by two separate nodes, the model building and its evaluation are actually performed in conjunction at the computing node when the chosen location is not local.

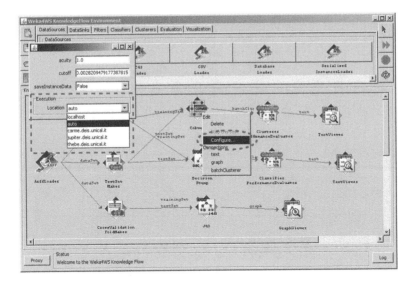

FIGURE 6.8   Weka4WS KnowledgeFlow: Selection of the computing node.

For complex workflow configurations, the subflow grouping feature of the KnowledgeFlow turns out to be useful in order to easily and quickly set the remote hosts for the execution of the algorithms. Through this feature it is possible to group together a set of workflow components that will then be represented graphically by only one component, like the black-to-gray faded one shown in Figure 6.9: right-clicking on this component makes it possible to either set to auto all the computing locations of the algorithms belonging to the group, or to choose the specific location of each algorithm by accessing the relative configuration listed in the menu.

The computations may be started, as shown in Figure 6.10, either by selecting the Start loading entry in the right-click context menu of each loader component of the flow (just like usually done in the conventional Weka KnowledgeFlow) or by pressing the Start-all-executions button in the right-top corner of the window (which is more convenient in flows with multiple loader components).

As for the Explorer component, pressing the Log button in the lower-right corner makes it possible to follow the computations in their single steps as well as to know their execution times.

6.4.1 Supporting Data-Parallel Workflows

The workflow shown above exploits independent parallelism, a form of parallelism that runs multiple independent tasks in parallel on different

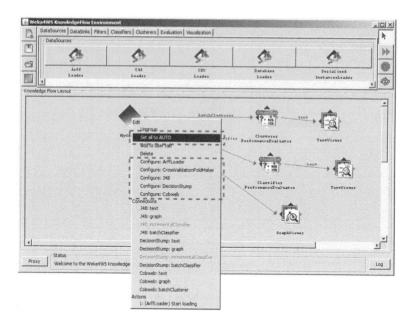

FIGURE 6.9   Weka4WS KnowledgeFlow: Computing node selection for a group of algorithms.

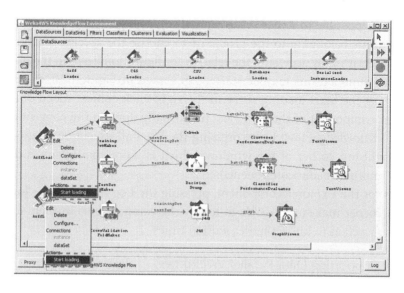

FIGURE 6.10   Weka4WS KnowledgeFlow: Computations start.

processors, available on a single parallel machine or on a set of machines connected through a network. Another form of parallelism that can be effectively exploited in data mining workflows is data parallelism (Pautasso and Alonso, 2006), where a dataset is split into smaller chunks, each chunk is processed in parallel, and the results of each processing are then combined to produce a single result.

Both forms of parallelism aim to achieve execution time speedup, and a better utilization of the computing resources, but while independent parallelism focuses on running multiple tasks in parallel so that the turnaround time is bound to the execution time of the slowest task, data parallelism focuses on reducing the execution time of a single task by splitting it into subtasks, each one operating on a subset of the original data.

Data parallelism is not natively supported in the KnowledgeFlow. However, it is easy to extend the Weka4WS KnowledgeFlow to also support data parallel applications. For example, in Lackovic, Talia, and Trunfio (2009b), the use of data parallelism in Weka4WS is demonstrated through a distributed classification workflow in which a dataset is partitioned into different subsets, which are analyzed in parallel by multiple instances of a given classification algorithm, and the resulting classifiers are then compared to choose the final model. To support this pattern, two new components, called *DataSetPartitioner* and *ModelChooser*, have been developed and included in the Weka4WS KnowledgeFlow.

A *DataSetPartitioner* component receives one dataset in input, divides it into a number of partitions equal to the number of its outgoing arcs, and assigns each partition to one workflow node representing a classification algorithm. If the data mining algorithm is annotated to be executed on a remote computing node, the subset assigned to it will be transferred to that computing node as described in Section 6.5. In the *DataSetPartitioner* the output partitions have by default the same number of instances. However, the size of each partition, expressed as a percentage of the original dataset size, may be changed by the user through a configuration panel. A *ModelChooser* node receives a set of base classifiers, resulting from the classification tasks performed on the different partitions, and returns the "best" model based on the criterion specified by the user (e.g., the lowest error rate).

An example of workflow using these two components is shown in Figure 6.11. An *ArffLoader* node is used to load a dataset from a file. Then, a *DataSetPartitioner* node is used to generate five subsets of equal size

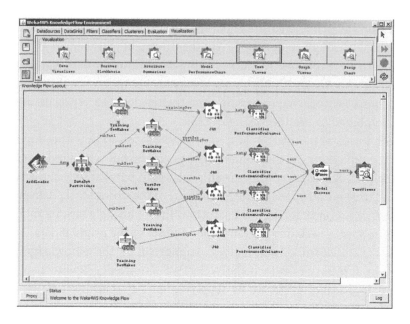

FIGURE 6.11   Weka4WS KnowledgeFlow: Example of data-parallel workflow.

from the incoming dataset. Each subset is sent to a different evaluation component, precisely to four *TrainingSetMakers* and one *TestSetMaker*.

Each one of the four *TrainingSetMakers* marks the incoming dataset partition as a training set and sends it to one of the four J48 classification algorithms, while the only *TestSetMaker* marks the incoming dataset partition as a test set and sends it to all four J48 algorithms. These are in turn connected with four *ClassifierPerformanceEvaluators*, which are all linked with one *ModelChooser* component. Finally, the *ModelChooser* selects the best model and sends it to a *TextViewer* component for its visualization.

6.5 EXECUTION MANAGEMENT

We conclude the chapter by describing in detail, through an invocation example, all the steps and mechanisms involved in the execution of a single data mining task in Weka4WS; these steps are the same regardless of whether the task is invoked from the Explorer or the KnowledgeFlow interface.

In this example we are assuming that the client is requesting the execution of a classification task on a dataset that is present in the user node, but not in the computing node where it will be analyzed. This is to be considered a worst-case scenario, because in many cases the dataset to be

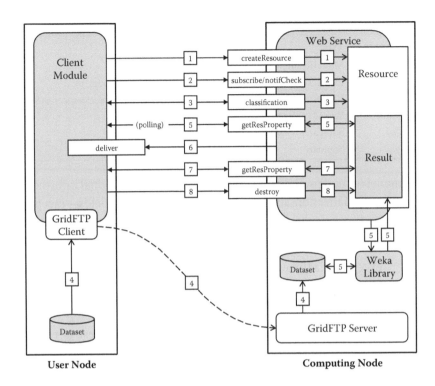

FIGURE 6.12   Execution steps of a data mining task in Weka4WS.

mined is already available (or replicated) on the computing node where the task is submitted.

When a remote data mining task is started, an unambiguous identification number, called *taskID*, is generated: this number is associated with that particular task and is used when a *stopTask* operation is invoked to identify the task among all the others running at the computing node.

The whole execution process may be divided into the eight steps shown in Figure 6.12:

1. *Resource creation*: The *createResource* operation is invoked to create a new resource that will maintain its state throughout all the subsequent invocations of the Web service until its destruction. The state will be stored as resource properties; more specifically a Result property, whose structure is detailed in Table 6.4, is used to store the results of the data mining task. The first three property fields store the inferred

TABLE 6.4   Structure of the Result Resource Property

Field Name	Type	Subfield Names
model	ModelResult	model, models
evaluation	EvalResult	summary, classDetails, matrix
visualization	VisResult	predInstances, predictions, plotShape, plotSize
exception	Weka4WSException	thrown, message
stopped	boolean	
ready	boolean	

model or models, the evaluation outcomes, and additional information about visualization and prediction. The other three fields—*exception*, *stopped*, and *ready*—are used only when certain circumstances arise:

- If during the computation phase something goes wrong and an exception is thrown, then the field *exception* is set accordingly: its boolean parameter *thrown* is set to true, and in its string parameter, *message*, the generated exception message is stored.

- If during the task execution a termination request is received through the *stopTask* operation, then the boolean field *stopped* is set to true.

- After the end of the computation the results are put into the Result property and the field *ready* is set to true: this field is used by the client when it is unable to receive notifications, like in the scenario depicted in Figure 6.2, to periodically check whether the results are ready.

After the resource has been created, the Web service returns the *endpoint reference* (EPR) of the resource created. The EPR is unique within the Web service and differentiates this resource from all the other resources in the same service. Subsequent requests from the client module will be directed to the resource identified by that EPR.

2. *Notification subscription and notifications check*: The *subscribe* operation is invoked in order to be notified about changes that will occur to the Result resource property. Upon these property value changes (that is, upon the conclusion of the data mining task), the client module will be notified of that. Just after the *subscribe* operation, the *notifCheck* operation is invoked to request the immediate

TABLE 6.5   Structure of a Response Object

Field Name	Type
datasetFound	boolean
testsetFound	boolean
dirPath	string
exception	Weka4WSException

delivery of a dummy notification to check whether the client is able to receive notifications, as described before: when and if the user node will receive this dummy notification, it will switch to the push-style mode; otherwise, it will persist in pull-style mode.

3. *Task submission*: The *classification* operation is invoked to submit the execution of a classification task. This operation requires the seven parameters shown in Table 6.2, among which is the *taskID* previously mentioned. The operation returns a Response object, whose structure is detailed in Table 6.5. If a copy of the dataset is not already available on the computing node, then the field *datasetFound* is set to false, and the *dirPath* field is set to the URL where the dataset has to be uploaded; similarly, when a validation is required on a test set that is different from the dataset, and the test set is not already available on the computing node, the *testsetFound* field is set to false. The URL where the test set has to be uploaded is the same as for the dataset. If during the invocation phase something goes wrong and an exception is thrown, then the *exception* field is set accordingly: its boolean parameter *thrown* is set to true, and its string parameter, *message*, will store the exception message generated.

4. *File transfer*: In this example we assumed that the dataset was not already available on the computing node, so the client module needs to transfer it to the computing node. To this end, a GridFTP client is instantiated and initialized to interoperate with the GridFTP server on the computing node: the dataset (or test set) is then transferred to the computing node machine and saved in the directory whose path was specified in the *dirPath* field contained in the Response object returned by the classification operation.

5. *Data mining*: The classification analysis is started by the Web service through the invocation of the appropriate Java class in the Weka library. The results of the computation are stored in the Result

property of the resource created in Step 1. In those cases where the client is unable to receive notifications (pull style), the client will be periodically checking the results for readiness through the value of the Result's field *ready* (see Table 6.4).

6. *Notification reception*: If the push-style notification mode is allowed, a notification message is sent to the client module by invoking its implicit *deliver* operation as soon as the Result property changes. This mechanism permits the asynchronous delivery of the execution results as soon as they are generated.

7. *Results retrieving*: The client module invokes the *getResourceProperty* operation to retrieve the Result property containing computation results.

8. *Resource destruction*: The client module invokes the *destroy* operation, which eliminates the resource created in Step 1.

CHAPTER 7

How Services Can Support Mobile Data Mining

T HIS CHAPTER DISCUSSES EXECUTION of pervasive data mining tasks from mobile devices through the use of Web services. Section 7.1 introduces mobile data mining. Section 7.2 discusses the use of Web services in mobile environments. Section 7.3 describes a client-server system for mobile data mining based on Web services. Finally, Section 7.4 presents a mobile-to-mobile (M2M) data mining architecture designed to enable mobile knowledge discovery in a wide range of wireless network scenarios.

7.1 MOBILE DATA MINING

The dissemination and growing power of wireless devices opened the way for running analysis and mining of data in mobile scenarios. Enabling *mobile data mining* is a significant added value for nomadic users, enterprises, and organizations that need to perform analysis of data generated either from a mobile device (e.g., sensor readings) or from remote sources.

A growing number of mobile-phone-based and personal digital assistant (PDA)–based data-intensive applications is appearing. Examples include cell-phone-based systems for body-health monitoring, vehicle monitoring, and wireless security systems. Monitoring data in small embedded devices for smart appliances, and onboard monitoring using nanoscale devices, are examples of applications that we may see in the near future. Support for advanced data mining is necessary for such mobile applications.

Mobile data mining has to face the typical issues of distributed data mining environments, and additional technological constraints such as low-bandwidth networks, relatively small storage space, limited availability of battery power, slower processors, and small displays to visualize the results (Pittie, Kargupta, and Park, 2003).

Mobile data mining may include different scenarios in which a mobile device can play the role of data producer, data analyzer, client of remote data miners, or a combination. More specifically, we can envision five basic scenarios for mobile data mining:

- A mobile device is used as a terminal for ubiquitous access to a remote server that provides some data mining services. In this scenario, the server analyzes data stored in a local or a distributed database and delivers the results of the data mining task to the mobile device for its visualization.

- Data generated in a mobile context are gathered through a mobile device and sent in a stream to a remote server to be stored into a local database. Data may be periodically analyzed by using specific data mining algorithms and the results used for making decisions about a given purpose.

- Mobile devices are used to perform data mining analysis. Due to the limited computing power and memory/storage space of today's mobile devices, it is not possible to execute heavyweight data mining tasks on such devices. However, some steps of the knowledge discovery in databases (KDD) process (e.g., data selection and preprocessing) or lightweight data mining algorithms that analyze small datasets can be effectively executed on mobile devices.

- A mobile device acts as a data mining server for other mobile clients. As stated earlier, data analysis provided by a mobile device may include either lightweight data mining algorithms or some steps of the KDD process.

- A mobile device acts as a gateway for other mobile devices. In this case, even if the mobile gateway does not provide processing, it plays the fundamental role of linking poorly connected devices to remote processing nodes.

In the remainder of this section, we discuss some significant examples of mobile data mining systems described in the literature.

7.1.1 Mobile Data Mining Systems Examples

MobiMine (Kargupta et al., 2002) is an example of a data mining environment designed for intelligent monitoring of stock markets from mobile devices. It is based on a client-server architecture. The clients, running on mobile devices such as PDAs, monitor a stream of financial data coming through a server. The server collects the stock market data from different Web sources in a database and processes it on a regular basis using several data mining techniques.

The clients query the database for the latest information about quotes and other information. A proxy is used for communication among clients and the database. Thus, when a user has to query the database, he or she sends the query to the proxy that connects to the database, retrieves the results, and delivers them to the client. To efficiently communicate data mining models over wireless links with limited bandwidth, MobiMine uses a Fourier-based approach to represent the decision trees, which saves both memory on mobile devices and network bandwidth.

Another example of a mobile data mining system was proposed by Wang, Helian, Guo, and Jin (2003). That system considers a single logical database that is split into a number of fragments. Each fragment is stored on one or more computers connected by a communication network, either wired or wireless. Each site is capable of processing user requests that require access to local or remote data.

Users can access corporate data from their mobile devices. Depending on the particular requirements of mobile applications, in some cases the user of a mobile device may log on to a corporate database server and work with data there. In other cases the user may download data and work with it on a mobile device or upload data captured at the remote site to the corporate database. The system defines a distributed algorithm for global association rule mining, which does not need to ship all of the local data to one site, thereby not causing excessive network communication cost.

Another interesting application of mobile data mining is the analysis of streams of data generated from mobile devices. Some possible scenarios are patient health monitoring, environment surveillance, and sensor networks. The VEhicle DAta Stream mining (VEDAS) system proposed by Kargupta et al. (2003) is an example of a mobile environment for monitoring and mining vehicle data streams in real time. The system is designed to monitor vehicles using on-board PDA-based systems connected through wireless networks.

VEDAS continuously analyzes the data generated by the sensors located on most modern vehicles, identifies the emerging patterns, and reports them to a remote control center over a low-bandwidth wireless network connection. The overall objective of VEDAS is to support drivers by characterizing their status, and to help the fleet managers by quickly detecting security threats and vehicle problems.

7.2 MOBILE WEB SERVICES

Growing interest in the use of Web services in mobile environments has been registered in the last decade. *Mobile Web services* make it possible to integrate mobile devices with server applications running on different platforms, allowing users to access and compose a variety of distributed services from their personal devices. Indeed, as in standard wired scenarios, Web services can be exploited in mobile environments to improve interoperability between clients and server applications independent from the different platforms they run on.

Basically, there are three architecture models for implementing Web services in mobile environments (Adaçal and Bener, 2006):

- Wireless Portal Network

- Wireless Extended Internet

- Peer-to-Peer (P2P) Network

In a Wireless Portal Network there is a gateway between the mobile client and the Web service provider. The gateway receives the client requests and takes care of issuing corresponding SOAP requests and returning responses in a specific format supported by the mobile device.

In the Wireless Extended Internet architecture, mobile clients interact directly with the Web service provider. In this case mobile clients are true Web service clients and can send or receive SOAP messages.

Finally, in a P2P Network, mobile devices can act as both Web service clients and providers. This capability of acting both as consumer and provider can be particularly useful in systems such as ad hoc networks. It represents the more general model that can offer very interesting opportunities for mobile services in the near future.

However, in most application scenarios, mobile devices act only as Web service consumers. In these cases, the choice between the Wireless Portal

Network and the Wireless Extended Internet architecture mainly depends on the level of performance required by the application.

The Wireless Extended Internet configuration requires mobile devices with eXtensible Markup Language (XML)/SOAP processing capabilities. This introduces additional processing load on the device and some traffic overhead for transporting SOAP messages over the wireless network (Tian et al., 2004). Although the additional processing load could be negligible in most devices, the traffic overhead can affect response time in the presence of wireless connections with limited bandwidth.

On the other hand, the Wireless Portal Network architecture requires the intermediation of a gateway that acts as a proxy between client requests and service providers. This enables the use of a set of optimizations (e.g., data compression, binary encodings) for reducing the amount of data transferred over the wireless link, but these methods generally depend on the specific structure of data used by the application (Adaçal and Bener, 2006), so their applicability is limited.

7.2.1 Mobile Web Services Initiatives

Following either the Wireless Portal Network or the Wireless Extended Internet architecture, some researchers studied how to improve functionalities and performance of Web services in mobile environments.

Adaçal and Bener (2006) proposed an architecture that includes the three standard Web service roles (provider, registry, and client) and three new components: a service broker, a workflow engine, and a mobile Web service agent. The mobile Web service agent acts as a gateway to Web services for mobile devices and manages all communications among mobile devices and the service broker or the workflow engine. The agent, which is located inside the mobile network, receives the input parameters required for service execution from the mobile device and returns the executed service. It also selects services according to user preferences and context information such as location, air-link capacity, or access-network type.

Chu, You, and Teng (2004) proposed an architecture that divides the application components into two groups: local components, which are executed on the mobile device, and remote components, which are executed on the server side. The system is able to dynamically reconfigure application components for local or remote execution to optimize a utility function derived from the user preferences. This approach implements a *smart client* model, which is in contrast with that of a *thin client* (which is only capable of rendering a user interface) generally implemented in wired scenarios.

Zahreddine and Mahmoud (2005) proposed an approach for Web service composition in which an agent performs the composition on behalf of a mobile user. In the proposed architecture, the client request is sent to a server that creates an agent on behalf of the user. The request is then translated into a workflow to be performed by the agent. The agent looks for services that are published in a Universal Description, Discovery, and Integration (UDDI) registry, thus retrieving the locations of multiple services that suit the request requirements. The agent then creates a specific workflow to follow, which entails the agent traveling from one platform to another, completing the tasks in the workflow.

Besides these and other research works on architectural aspects, some companies worked on the implementation of a software library, named JSR-172, which provides standard access to Web services from mobile devices.* JSR-172 is available as an additional library for the Java 2 Micro Edition (J2ME) platform†; thus, it can be used on mobile devices that support the Java technology.

The main goal of JSR-172 is to enable interoperability of J2ME clients with Web services. It does so by providing

- Application programming interfaces (APIs) for basic manipulation of structured XML data, based on a subset of standard APIs for XML parsing.
- APIs and conventions for enabling XML-based remote procedure call (RPC) communication from J2ME, including the definition of a strict subset of the standard Web Service Description Language (WSDL)-to-Java mapping, suitable for J2ME; stub APIs based on this mapping for XML-based RPC communication; and runtime APIs to support stubs generated according to the mapping above.

The mobile data mining system described in the next section implements a wireless extended Internet architecture and uses the JSR-172 library for the implementation of its client application.

7.3 SYSTEM FOR MOBILE DATA MINING THROUGH WEB SERVICES

The goal of the system presented in Talia and Trunfio (2010b) is supporting mobile data mining operations from small mobile devices through

---

* Java Community Process, JSR 172: J2ME™ Web Services Specification, http://jcp.org/en/jsr/detail?id=172 (accessed June 2012).

† ORACLE, Java ME and Java Card Technology, http://java.sun.com/javame (accessed June 2012).

the use of Web services. In the following sections we describe the system architecture, the client and server software components, the basic execution mechanisms, and the system implementation.

7.3.1 System Architecture

The system is based on the simple client-server architecture shown in Figure 7.1. The architecture includes three types of components:

- *Data providers*: Applications that generate the data to be mined.

- *Mobile clients*: Applications that require the execution of data mining tasks on remote data.

- *Mining servers*: Server nodes used for storing the data generated by data providers and for executing the data mining tasks submitted by mobile clients.

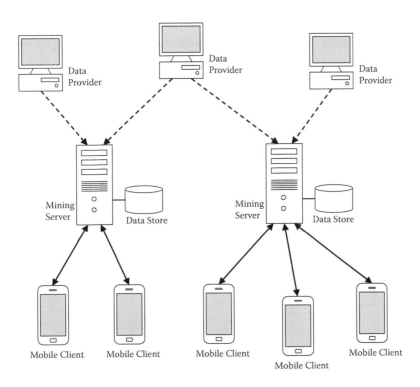

FIGURE 7.1   A client-server architecture for mobile data mining.

As shown in Figure 7.1, data generated by data providers are collected by a set of mining servers that store them in local data stores. Depending on the application requirements, data coming from a certain provider could be stored by multiple mining servers.

The main role of mining servers is to allow mobile clients to perform data mining on remote data by invoking a predefined set of data mining algorithms. Once connected to a given server, the mobile client allows a user to select the remote dataset to be analyzed and the data mining algorithm to be run. When a data mining task has been completed on a mining server, the task results are visualized on the user device either in textual or visual form.

7.3.2 Mining Server Components

Each mining server exposes its functionalities through two Web services: the *Data Collection Service* (*DCS*) and the *Data Mining Service* (*DMS*). Figure 7.2 shows the DCS, DMS, and the other software components of a mining server.

The DCS is invoked by data providers to store data on the server. The DCS interface defines a set of basic operations for uploading a new dataset, updating an existing dataset with incremental data, or deleting an existing dataset. These operations are cumulatively indicated as *DCS ops* in the figure. As shown in the figure, the DCS performs either store or update operations on the local datasets in response to data providers requests.

The DMS is invoked by mobile clients to submit data mining tasks. Its interface defines a set of operations (*DMS ops*) that allow one to obtain the

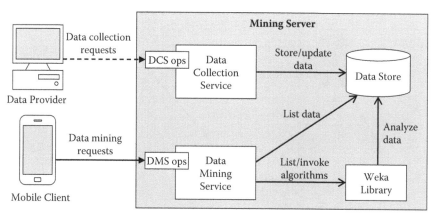

FIGURE 7.2   Software components of a mining server.

TABLE 7.1   Main Operations Provided by the Data Mining Service (DMS)

Operation	Description
listDatasets	Returns the list of datasets available.
listAlgorithms	Returns the list of data mining algorithms available.
submitTask	Submits a data mining task for the analysis of a given dataset using a specified algorithm; returns a unique task id.
getStatus	Returns the status of the task with a given id; the task status can be running, done, or failed.
getResult	Returns the result of the task with a given id, either in textual or visual form.

list of the available datasets and algorithms, submit a data mining task, get the current status of a given task, and get the result of a given task. Table 7.1 lists and describes the main operations provided by the DMS.

The data analysis is performed by the DMS using a subset of the algorithms provided by the Weka library (Hall et al., 2009). When a data mining task is submitted to the DMS, the appropriate algorithm of the Weka library is invoked to analyze the local dataset specified by part of the mobile client request.

7.3.3 Mobile Client Components

The mobile client is composed of three components: the *MIDlet*, the *DMS Stub*, and the *Record Management System* (*RMS*) (see Figure 7.3).

The MIDlet is a J2ME application allowing a user to submit data mining tasks and visualize their results. The DMS Stub is a Web service stub allowing the MIDlet to invoke the operations of a remote DMS. The stub is generated from the DMS interface to conform with the JSR–172 specifications. Even if, from a logical point of view, the DMS Stub and the MIDlet

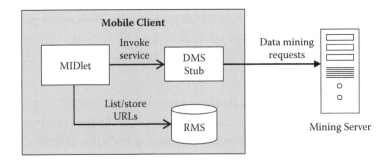

FIGURE 7.3   Software components of a mobile client.

are two separated components, they are distributed and installed as a single J2ME application.

The RMS is a simple record-oriented database that allows J2ME applications to persistently store data across multiple invocations. In this system, the MIDlet uses the RMS to store the URLs of the remote DMSs that can be invoked by the user. The list of URLs stored in the RMS can be updated by the user using a MIDlet functionality.

7.3.4 Execution Mechanisms

We describe here the typical steps that are executed by the client and server components to execute a data mining task in the system:

1. The user starts the MIDlet on his or her mobile device. After started, the MIDlet accesses the RMS and gets the list of remote mining servers. The list is presented to the user who selects the mining server to connect with.

2. The MIDlet invokes the *listDatasets* and *listAlgorithms* operations of the remote DMS in order to get the lists of datasets and algorithms that are available on that server. The lists are presented to the user who selects the dataset to be analyzed and the mining algorithm to be run.

3. The MIDlet invokes the *submitTask* operation of the remote DMS, passing the dataset and the algorithm selected by the user with associated parameters. The task is submitted in a batch mode: as soon as the task has been submitted, the DMS returns a unique *id* for it, and the connection between client and server is released.

4. After task submission, the MIDlet monitors its status by querying the DMS. To this end, the MIDlet periodically invokes the *getStatus* operation, which receives the task *id* and returns its current status (see Table 7.1). The polling interval is an application parameter that can be set by the user.

5. As soon as the *getStatus* operation returns *done*, the MIDlet invokes the *getResult* operation to receive the result of the data mining analysis. Depending on the type of data mining task, the MIDlet asks the user how to visualize the result of the computation (e.g., pruned tree, confusion matrix, etc.).

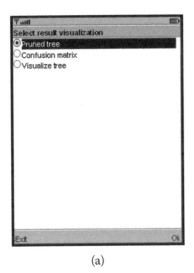 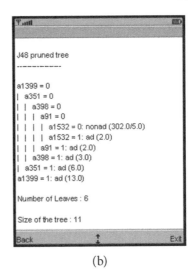

(a)   (b)

FIGURE 7.4   Two screenshots of the client application running on the emulator of the Sun Java Wireless Toolkit: (a) selecting which part of the result must be visualized; (b) visualizing the selected classification tree.

7.3.5 System Implementation

All the system components, but the Data Collection Service, have been implemented (Talia and Trunfio, 2010b). The mobile client was developed using the Sun Java Wireless Toolkit,* which is a widely adopted suite for the development of J2ME applications.

As mentioned earlier in this chapter, the small display size is one of the main limitations of mobile device applications. In data mining tasks, in particular, a small display size can affect the appropriate visualization of complex results representing the discovered model. The system overcomes this limitation by splitting the result into different parts and allowing users to select which part to visualize at one time.

Moreover, a user can choose to visualize the mining model (e.g., a cluster assignment or a decision tree) either in textual form or as an image. In both cases, if the information does not fit the display size, the user can move on it by using the normal navigation facilities of the mobile device.

As an example, Figure 7.4 shows two screenshots of the mobile client taken from a test application. In this example, the MIDlet is executed on

---

* ORACLE, Sun Java Wireless Toolkit for CLDC, http://java.sun.com/products/sjwtoolkit (accessed June 2012).

the emulator of the Sun Java Wireless Toolkit, while the Data Mining Service is executed on a remote server. The screenshot on the left shows the menu for selecting which part of the result of a classification task must be visualized, while the screenshot on the right shows the result (in this case, the pruned tree resulting from classification).

The system has been tested using some datasets from the University of California, Irvine (UCI) machine learning repository,* and some data mining algorithms provided by the Weka library. The experiments showed that the system performance depends almost entirely on the computing power of the server on which the data mining task is executed. On the contrary, the overhead due to the communication between MIDlet and Data Mining Service does not affect the execution time in a significant way, because the amount of data exchanged between client and server is very small. In general, when the data mining task is relatively time consuming, the communication overhead is a negligible percentage of the overall execution time.

7.4 MOBILE-TO-MOBILE (M2M) DATA MINING ARCHITECTURE

The system presented in the previous section is based on a client-server architecture in which a mobile device is used as a terminal for ubiquitous access to remote data mining servers. In general, however, mobile devices may play the role of data producer, data miner, or gateway for other mobile devices, as already discussed in Section 7.1.

To cope with the wide variety of scenarios that can be envisioned in mobile data mining applications, a research project called *M2M Data Mining* was started at the University of Calabria† with the goal of exploiting *mobile-to-mobile* (*M2M*) technologies to support pervasive and ubiquitous data mining through mobile devices. The project builds around a M2M architecture providing services for distributed wireless data mining algorithms and applications.

A typical M2M Data Mining scenario includes both *stationary nodes* (e.g., computer servers) and *mobile devices* (e.g., mobile phones, PDAs), as shown in Figure 7.5.

The mobile devices interact with each other in a peer-to-peer manner to deal with processing power and energy capacity limitations. Whenever a

---

* UCI, Machine Learning Repository, http://archive.ics.uci.edu/ml (accessed June 2012).
† Grid Computing Lab, M2M Data Mining: Overview, http://grid.deis.unical.it/m2m-dm (accessed June 2012).

resource-limited computing device in such a cooperative environment has a set of tasks (or subtasks) to be executed (which may have interdependencies and intercommunication requirements), it may exploit the computing resources provided by nearby mobile devices or by a remote stationary node.

To ease the cooperation between mobile devices, they can be grouped into *clusters* also referred to as *mobile groups*. Clusters may be formed on the basis of various criteria such as transmission range, number and type of mobile devices, and their geographical location. Each cluster has a node referred to as the *cluster-head*, which is the mobile group coordinator and acts as a gateway between ordinary cluster nodes and neighboring mobile groups.

Mobile devices within a cluster interact with each other through ad hoc connections (e.g., bluetooth, Wi-Fi) referred to as *M2M interactions*, which are represented as dashed arrows in Figure 7.5. Connections between mobile groups (*cluster-to-cluster interactions*) take place through ad hoc links between the cluster-heads and are represented as dotted-dashed arrows in the figure. Mobile groups can be connected to stationary nodes

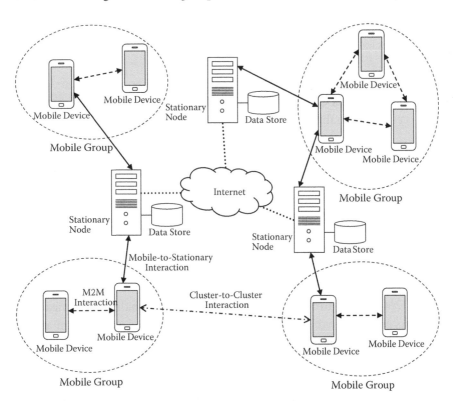

FIGURE 7.5 The mobile-to-mobile (M2M) data mining architecture.

through their cluster-heads (*mobile-to-stationary interactions*) by exploiting an Internet connection (e.g., Wi-Fi, WiMAX). Stationary nodes may be connected to each other through a wired or a wireless Internet connection.

In the M2M data mining architecture, stationary nodes and mobile devices provide a set of functionalities exposed as services. The functionalities provided by stationary nodes can be broadly classified into three groups:

- *Knowledge discovery*: Services to execute or support the different steps of the knowledge discovery process (preprocessing, data mining, visualization, etc.).

- *Data management*: Services allowing storage and retrieval of data (e.g., data generated either by mobile devices or by third-party data providers).

- *Coordination*: Services allowing mobile devices to organize themselves into mobile groups and to manage computations in a cooperative way (e.g., registration services, discovery services, etc.).

Mobile devices, in turn, provide the following functionalities:

- *M2M knowledge discovery*: Services for running knowledge discovery tasks that can be executed on limited resources, such as preprocessing, visualization, or lightweight data mining processes.

- *M2M resource management*: Services allowing local resources (e.g., memory, CPU load, battery status) to be monitored to establish whether the device is able to execute a given data mining task.

- *M2M coordination*: Services enabling mobile devices to organize themselves into mobile groups on a temporary basis for on-purpose knowledge discovery applications.

7.4.1 M2M Data Mining Implementation

The M2M data mining architecture and its functionalities are meant to serve as a reference model that can be specialized and extended in actual implementations on the basis of system goals and application scenarios.

An implementation of the M2M data mining architecture, focusing on energy efficiency, was proposed in Comito, Talia, and Trunfio (2010). Ensuring energy efficiency is a key aspect to be addressed to enable effective and reliable data mining over mobile devices, because most commercially available mobile devices like PDAs and mobile phones have battery power that would last for only a few hours.

As a first step, the energy-efficient M2M Data Mining initiative concentrated on implementing a set of specialized software components for mobile devices that cooperatively perform the functionalities introduced in the previous section, with particular focus on energy efficiency. As shown in Figure 7.6, each mobile device includes four software components: *Resource Information Service* (RIS), *M2M Coordination Service* (MCS), *Energy-Aware Scheduler* (EAS), and *Knowledge Discovery Service* (KDS).

The RIS is responsible for collecting information about the resources of a mobile device and the context in which an application is running, in order to adapt its execution. To this aim, the RIS is composed of two modules:

- *Resource Monitoring* module: It provides dynamic information about mobile device resources such as CPU utilization, energy consumption, battery level, available memory, and network connectivity performance.

- *Resource Evaluator* module: This module acts as a resource measurement receiver from both local and environmental resources. It then takes appropriate actions on the basis of the measures received (e.g., choosing the most suitable configuration for the execution of a given data mining task). Moreover, the module is responsible for starting the data mining task with the appropriate parameters.

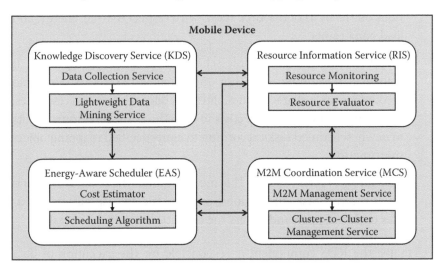

FIGURE 7.6 Software components of mobile devices for energy-efficient mobile-to-mobile (M2M) data mining.

The MCS is responsible for the coordination of mobile devices and includes two modules:

- *Mobile-to-Mobile Management* module: It implements mechanisms to support the coordination of nodes within a mobile group such as cluster formation and maintenance, joining and leaving a cluster, cluster-head election and replacement, and cooperative data mining tasks execution.

- *Cluster-to-Cluster Management* module: This module provides mechanisms for cluster-to-cluster interactions, for allocating data mining tasks, and for supporting their collaborative execution across different mobile groups.

The EAS is responsible for task assignment among clusters. It implements a scheduling strategy aimed at prolonging network lifetime by distributing the energy consumption between the mobile groups. The scheduler interacts with the RIS component through its resource monitoring and resource-evaluator modules. The scheduler is also tightly related to the KDS component, because it is actually the scheduler that activates a data mining process. The EAS includes three modules:

- *Cost Estimator* module: This module exploits information about availability, performance, and cost of resources collected by the RIS component. It deals with the computation of the estimation functions on the basis of resource status with respect to time, energy, and load constraints.

- *Mapper* module: This module schedules the data mining tasks. It embeds a scheduling algorithm and a matchmaker that takes into account resource characteristics, interdependencies between resources, and computational and I/O costs to map the available resource units to newly scheduled tasks according to a prespecified mapping objective function.

- *Scheduling Process* module: This module guides the scheduling activity. It receives task submissions, requests the corresponding schedules to the mapper, and orders the execution of scheduled tasks.

Finally, the KDS component is responsible for the execution of data mining tasks over a mobile device. It includes two modules:

- *Data Collection* module: This module provides access/store mechanisms for data to be processed or generated as a result of a data mining process. Typically, only a limited amount of data can be stored on a mobile device. Therefore, this module will also manage the interactions with stationary nodes acting as data storage servers or data sources.

- *Lightweight Data Mining* module: It is responsible for managing the execution of a data mining task on the mobile device. If the local resources are not (or no more) sufficient to carry out the computation, this module can delegate the process to another node. As an example, this may happen when the resource monitoring process indicates that the device cannot ensure the desired level of accuracy according to the incoming data rate.

Early evaluation of the system, focusing in particular on the cluster formation mechanisms, was recently presented by Comito, Talia, and Trunfio (2011). A combined weighted metric approach was used to select cluster-head nodes, taking into account energy, mobility, and location of mobile nodes. The clustering scheme was evaluated using a prototype of the system that included smart phones and Android emulators. The experimental results showed that using a clustering approach results in high average residual life of mobile devices, thus proving its effectiveness as a strategy for prolonging network lifetime.

CHAPTER 8

Knowledge Discovery Applications

THIS CHAPTER DISCUSSES SOME EXAMPLES of distributed knowledge discovery applications implemented in the Knowledge Grid and the Weka4WS frameworks as workflows of services. The examples discussed in this chapter show how a set of distributed data mining tasks are developed and run as a collection of services cooperating according to the service-oriented architecture (SOA) model. The applications presented here demonstrate the effectiveness of the service-oriented approach for distributed knowledge discovery, as well as the performance improvements deriving from their execution on a Grid environment.

8.1 KNOWLEDGE GRID APPLICATIONS

The Knowledge Grid has been used to develop several knowledge discovery applications in different domains. Such applications have been useful for evaluating the overall system under different aspects, including performance.

A first example, from the bioinformatics domain, is an application that makes use of the Knowledge Grid to perform the clustering of a large collection of human protein sequences (Cannataro et al., 2004). The overall application is composed of four phases: data selection, data preprocessing, clustering, and results visualization. The clustering of protein sequences is performed by TribeMCL, a method that clusters correlated proteins into groups termed "protein families" (Enright, Dongen, and Ouzounis, 2002). The clustering is achieved by analyzing similar patterns between proteins in a given dataset, and by using these patterns to assign proteins to related groups. The application has been used to analyze all the human proteins in the SwissProt database,[*] a large collection of protein sequences obtained from the translation of DNA sequences or collected from the scientific literature or applications. The distributed execution of the application on a three-node Grid resulted in a turnaround time of about 12 hours, which, compared to the turnaround time obtained on a single node (27 hours), corresponds to an execution speedup of about 2.3.

Another significant example is the distributed execution of Knowledge Discovery in Databases Markup Language (KDDML)–based data mining applications on the Knowledge Grid (Bueti, Congiusta, and Talia, 2004). KDDML is an eXtensible Markup Language (XML)–based query language for performing knowledge extraction from databases (Alcamo, Domenichini, and Turini, 2000). A KDDML query has the structure of a tree in which each node is a KDDML operator specifying the execution of a KDD task (preprocessing, classification, etc.), or the logical combination (and/or operators) of results coming from lower levels of the tree. A query representing a KDDML process can be split into a set of subqueries. The Knowledge Grid has been used to distribute the subqueries to a set of Grid nodes for enabling their concurrent execution. After the query acquisition and its splitting into a set of subqueries, a set of Grid nodes are selected and both the query executor and the related subqueries are transferred to each of these nodes. Once the execution on a node is completed, the partial results are moved to another node for subsequent processing, according to the structure of the query tree. The execution of a KDDML application on a four-node Grid has shown that the benefit of using the Knowledge Grid increases with the dataset size. For instance, the extraction of a classification model from a dataset of 320,000 tuples, according to a two-step process (applying first a clustering algorithm and then a classification one), resulted in a speedup of about 3.7 compared to the sequential execution, reducing the turnaround time from about 3 hours to 45 minutes.

Other significant examples are presented in Congiusta, Talia, and Trunfio (2003, 2006). The former describes a Knowledge Grid application in which a set of banks cooperate to extract a loan-scoring prediction model, in the form of a decision tree, based on the information each of them owns about its clients. The application is composed of two main phases: the induction of the decision trees, performed locally at each bank;

---

[*] EMBL-EBI, UniProt: Welcome to UniProt, http://www.ebi.ac.uk/swissprot (accessed June 2012).

and the models' combination and validation, executed at a central site after the partial models have been produced and moved there. The latter describes how the Knowledge Grid can be used for general-purpose data-intensive applications. As an example, it is shown how a home videotape can be converted to a video CD in a distributed manner using the Knowledge Grid services, assuming that all the packages and libraries needed to perform acquisition, compression, and CD burning are available on a set of Grid nodes.

In the remainder of this section we focus in particular on the DIS3GNO system, showing how it has been used to design and run two distributed knowledge discovery workflows on the Knowledge Grid (Cesario, Lackovic, Talia, and Trunfio, 2012). The first workflow is a parameter sweeping application in which a dataset is processed using multiple instances of the same classification algorithm with different parameters, with the goal of finding the best classifier based on some accuracy parameters. The second workflow represents the ensemble learning application introduced in Section 5.3.2. Both of these workflows have been executed on a set of Grid nodes to evaluate their scalability.

8.1.1 Classification with Parameter Sweeping

In this application a dataset is analyzed by running multiple instances of the same classification algorithm, with the goal of obtaining multiple classification models from the same data source. The covertype* input dataset contains information about forest cover type for a large number of sites in the United States. Each dataset instance, corresponding to a site observation, is described by 54 attributes that give information about the main features of a site (e.g., elevation, aspect, slope, and so forth). The 55th attribute contains the covertype, represented as an integer in the range 1 to 7.

DIS3GNO has been used to design the application workflow shown in Figure 8.1, in which eight independent instances of the J48 classification algorithm perform a different classification task on the covertype dataset. In particular, each J48 instance is asked to classify data using a different value of confidence, ranging from 0.15 to 0.5.

The workflow, designed using the composition tools provided by DIS3GNO as described in Section 5.3.2, includes a dataset node (representing the covertype dataset) connected to eight tool nodes, each one associated

---

* UCI KDD Archive, Forest CoverType, http://kdd.ics.uci.edu/databases/covertype/covertype.html (accessed June 2012).

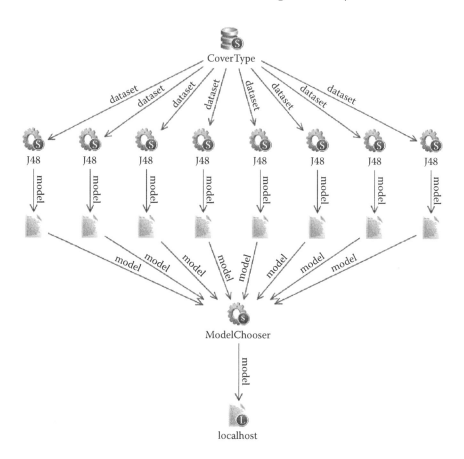

FIGURE 8.1   Classification with parameter sweeping: Screenshot of the workflow from the DIS3GNO graphical user interface (GUI).

with an instance of the J48 classification algorithm with a given value of confidence (ranging from 0.15 to 0.5). These nodes are in turn connected to another tool node, associated with a model chooser tool that selects the best classifier among those learned by the J48 instances. Finally, the node associated with the model chooser is connected to a model node having the location set to localhost; this enforces the model to be transferred to the client host for its visualization.

The same workflow has been executed multiple times on the Knowledge Grid, varying both the size of the input dataset and the number of computing nodes. This has enabled an evaluation of the speedup that can be achieved through the distributed execution of the workflow. The Grid used to execute this workflow included 11 nodes equipped with different

TABLE 8.1   Classification with Parameter Sweeping: Task Assignments
and Turnaround Times (hh:mm:ss)

Number of Nodes	Task Assignments (Node ← Tasks)	Turnaround Time 9 MB	Turnaround Time 18 MB	Turnaround Time 36 MB
1	$N_1 \leftarrow DM_1 \ldots DM_8$	2:43:47	7:03:46	20:36:23
2	$N_1 \leftarrow DM_1, DM_3, DM_5, DM_7$			
	$N_2 \leftarrow DM_2, DM_4, DM_6, DM_8$	1:55:19	4:51:24	14:14:40
4	$N_1 \leftarrow DM_1, DM_5$			
	$N_2 \leftarrow DM_2, DM_6$			
	$N_3 \leftarrow DM_3, DM_7$			
	$N_4 \leftarrow DM_4, DM_8$	58:30	2:26:48	7:08:16
8	$N_i \leftarrow DM_i$ for $1 \le i \le 8$	32:35	1:21:32	3:52:32

processors, having a computing power ranging from that of a Pentium IV with 2.4 GHz to that of a Xeon 5160 with 3 GHz, with a RAM size ranging from 2 to 4 GB.

Table 8.1 reports the turnaround times of the workflow obtained with three dataset sizes (9 MB, 18 MB, and 36 MB, which correspond, respectively, to 72,500, 145,000, and 290,000 instances of the original covertype dataset), when 1, 2, 4, and 8 computing nodes are used. The eight classification tasks that are included in the workflow are indicated as $DM_1 \ldots DM_8$, corresponding to the tasks of running J48 with a confidence value of 0.15, 0.20, 0.25, 0.30, 0.35, 0.40, 0.45, and 0.50, respectively. Table 8.1 shows how the classification tasks are assigned to the computing nodes (denoted as $N_1 \ldots N_8$), as well as the turnaround times for each dataset size.

When the workflow is executed on more than one node, the turnaround time includes the overhead due to file transfers. For example, in the Grid used to run this workflow, the transfer of a 36 MB dataset from the user node to a computing node takes, on average, 15 seconds. This value is small compared to the amount of time required to run a classification algorithm on the same dataset, which takes between 2.5 and 3.9 hours depending on the computing node. The turnaround time also includes the amount of time needed to invoke all the involved Knowledge Grid services (i.e., Execution Plan Management Service [EPMS], Resource Allocation and Execution Management Service [RAEMS], Knowledge Directory Service [KDS]) as required by the workflow. However, this amount of time is negligible compared to the turnaround time.

The turnaround times and speedup values for different number of nodes and dataset sizes are shown in Figure 8.2. For the 36 MB dataset, the

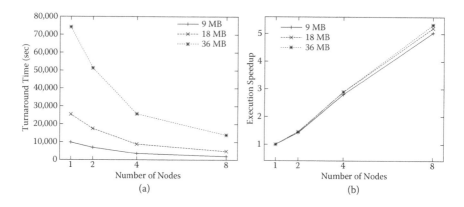

FIGURE 8.2 Classification with parameter sweeping: (a) turnaround times and (b) speedup values.

turnaround time decreases from more than 20 hours to less than 4 hours, passing from one to eight computing nodes. The execution speedup ranges from 1.45 using two nodes, to 5.32 using eight nodes. Similar trends have been registered with the other two dataset sizes, thus proving the good scalability of this application.

8.1.2 Ensemble Learning

The application discussed here is based on the ensemble learning scenario introduced in Section 5.3.2, whose corresponding workflow is reported in Figure 8.3 for the reader's convenience.

The input dataset, kddcup99,* contains a wide set of data produced during 7 weeks of monitoring in a military network environment subject to simulated intrusions. The workflow splits the dataset into two parts: a test set (one third of the original dataset) and a training set (two thirds of the original dataset). The latter is processed using four classification algorithms: ConjuctiveRule, NaiveBayes, RandomForest, and J48. The models generated by the four algorithms are then collected to a node where they are given to a voter component; the classification is performed and evaluated on the test set by taking a vote, for each instance, on the predictions made by each classifier.

As for the application described in the previous section, the ensemble learning workflow has been executed varying both the dataset size and the number of computing nodes, so as to evaluate the execution speedup.

---

* UCI KDD Archive, KDD Cup 1999 Data, http://kdd.ics.uci.edu/databases/kddcup99/kddcup99. html (accessed June 2012).

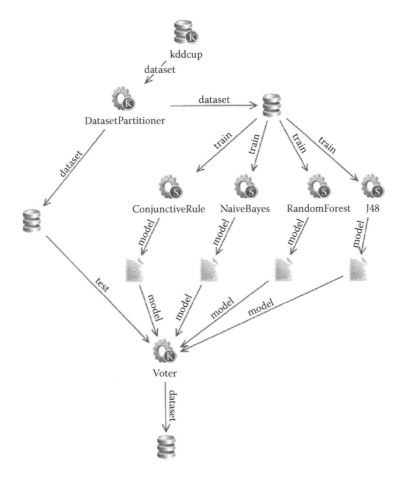

FIGURE 8.3 Ensemble learning: Screenshot of the workflow from the DIS3GNO graphical user interface (GUI).

Table 8.2 reports the turnaround times of the workflow obtained with three dataset sizes (100 MB, 140 MB, and 180 MB, corresponding to 940,000, 1,315,000, and 1,692,000 instances of kddcup99), using one, two, and four nodes. The four tasks are denoted as $DM_1...DM_4$, corresponding to ConjuctiveRule, NaiveBayes, RandomForest, and J48, respectively. The table shows also how the tasks are assigned to the computing nodes used for this application.

The turnaround times and speedup values for different number of nodes and dataset sizes are reported in Figure 8.4. In this case, the speedup is lower than that obtained with the application discussed in Section 8.1.1. This is due to the fact that the four algorithms require very different amounts of time to complete their execution on a given dataset. In fact, the

TABLE 8.2   Ensemble Learning: Task Assignments and Turnaround Times (hh:mm:ss)

Number of Nodes	Task Assignments (Node ← Tasks)	Turnaround Time 100 MB	Turnaround Time 140 MB	Turnaround Time 180 MB
1	$N_1 \leftarrow DM_1...DM_4$	1:30:50	2:31:14	3:34:27
2	$N_1 \leftarrow DM_1, DM_3$			
	$N_2 \leftarrow DM_2, DM_4$	1:03:47	1:37:05	2:07:05
4	$N_i \leftarrow DM_i$ for $1 \leq i \leq 4$	46:16	1:13:47	1:37:23

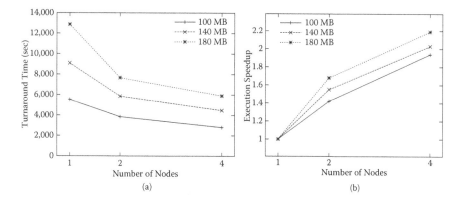

FIGURE 8.4   Ensemble learning: (a) turnaround times and (b) speedup values.

turnaround time is bound to the execution time of the slowest algorithm, thus limiting the speedup. However, the absolute amount of time saved by running the application on a distributed environment is still significant, particularly for the largest dataset when four computing nodes are used.

8.2 WEKA4WS APPLICATIONS

As described in Chapter 6, Weka4WS provides two user interfaces: Explorer and KnowledgeFlow. While the Explorer can be used to run applications composed of a single data mining task, the KnowledgeFlow allows users to run knowledge discovery applications in which multiple tasks are combined into a workflow. Here, we focus on the KnowledgeFlow interface, showing how it has been used to run two distributed data mining workflows on a Grid (Lackovic, Talia, and Trunfio, 2009a). The first workflow defines a classification application in which a dataset is analyzed using multiple algorithms, while the second workflow defines a clustering application using parameter sweeping. In addition, we discuss the execution of a workflow on a multicore machine to show how Weka4WS can obtain lower execution times compared to Weka even when it is executed on a single computer.

8.2.1 Classification Using Multiple Algorithms

The workflow described here has been designed through the Weka4WS KnowledgeFlow environment. It performs the analysis of the kddcup99 dataset using four different classification algorithms implemented as services: DecisionStump, NaiveBayes, J48, and RandomForest. From the original dataset all but nine attributes and the class attribute were removed using a selection filter provided by the Preprocess panel of the Explorer environment. Then, from the resulting dataset, another filter was used to extract three datasets with 215,000, 430,000, and 860,000 instances, and sizes of about 7.5 MB, 15 MB, and 30 MB. The workflow has been executed considering each one of these datasets as input.

The workflow, shown in Figure 8.5, begins with an *ArffLoader* node, used to load the kddcup99 dataset from the file, which is connected to a *CrossValidationFoldMaker* node (set to five folds), used for splitting the dataset into training and test sets according to a cross-validation. The *CrossValidationFoldMaker* node is connected to four nodes, each one executing the four algorithms mentioned earlier. These are in turn connected

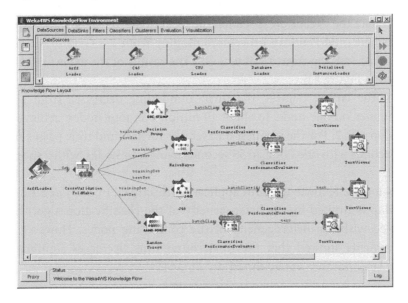

FIGURE 8.5 Classification using multiple algorithms: Screenshot of the workflow from the Weka4WS KnowledgeFlow.

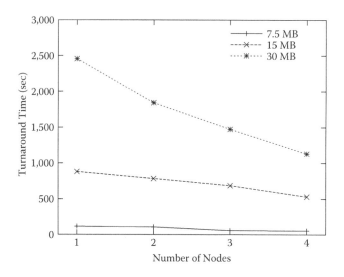

FIGURE 8.6   Classification using multiple algorithms: Turnaround times.

to a *ClassifierPerformanceEvaluator* node for the model validation, and then to a *TextViewer* node for results visualization.

When the user starts the application, the four branches of the workflow are executed in parallel on different Grid nodes. For each one of the three dataset sizes, the workflow has been executed using from one to four nodes to evaluate the execution speedup. The Grid nodes used for these experiments had Intel Pentium processors ranging from 2.8 GHz to 3.2 GHz, and RAM ranging from 1 GB to 2 GB. The results of the experiments are shown in Figure 8.6.

For the largest dataset, the turnaround time decreases from about 41 minutes on one node, to about 19 minutes on four nodes, achieving a speedup of 2.2. For the smallest dataset, the turnaround time passes from 117 seconds with one node, to 55 seconds using four nodes, with a speedup of 2.1.

As for the Knowledge Grid application discussed in Section 8.1.2, the speedup is limited by the heterogeneity of the classification algorithms, coupled with the heterogeneity of the computing nodes. However, the experiments showed that the turnaround time decreases significantly when large datasets are analyzed.

8.2.2 Clustering with Parameter Sweeping

In this example the Weka4WS KnowledgeFlow has been used to compose a parameter sweeping application in which the covertype dataset is analyzed

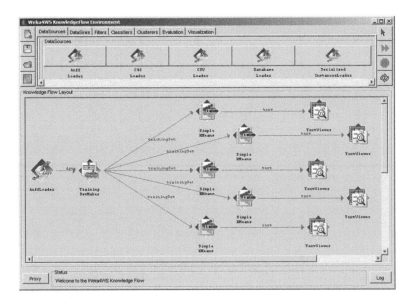

FIGURE 8.7   Clustering with parameter sweeping: Screenshot of the workflow from the Weka4WS KnowledgeFlow.

by running five parallel instances of the K-means algorithm with the goal of obtaining multiple clustering models from the same data source. The workflow corresponding to this application is shown in Figure 8.7.

This workflow includes an *ArffLoader* node connected to a *TrainingSetMaker* node used for accepting a dataset and producing a training set. The *TrainingSetMaker* node is connected to five nodes, each one performing the K-means clustering algorithm, and each one configured to group data into a different number of clusters (three to seven), based on all the attributes but the last one (the covertype). These nodes are in turn connected to a *TextViewer* node for results visualization.

The workflow has been executed using a number of computing nodes ranging from one to five, and dataset sizes of 9 MB, 18 MB, and 36 MB. The turnaround times are shown in Figure 8.8.

With the largest dataset as input, the turnaround time decreases from about 22 minutes obtained using one computing node, to 10.5 minutes obtained with five nodes. For the smallest dataset, the turnaround time decreases from 403 to 175 seconds passing from one to five nodes. The execution speedup with five nodes ranged from 2.1 to 2.8.

Different from the parameter sweeping application described in Section 8.1.1, in which the homogeneity of the classification tasks in terms

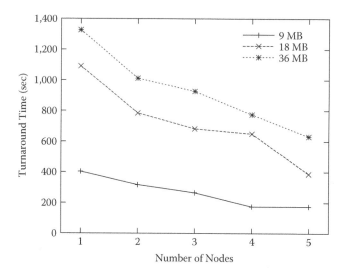

FIGURE 8.8   Clustering with parameter sweeping: Turnaround times.

of execution time led to high speedup values, in this case the speedup is limited by the different amounts of time taken by the various clustering tasks included in the workflow, because the slower task (i.e., the one in charge of grouping data into seven clusters) determines the turnaround time.

8.2.3 Local Execution on a Multicore Machine

Because Weka4WS executes an independent thread for each branch of the workflow, we obtain lower turnaround times compared to Weka even when the workflow is run on a single multiprocessor or multicore machine. To show this characteristic, we present in this section the turnaround times of a classification workflow when it is executed locally on a multicore machine.

The workflow considered here is a small-scale version of the parameter sweeping workflow presented in Section 8.1.1. In this case, the covertype dataset is analyzed in parallel using four instances of the J48 classification algorithm, configured to use a confidence factor of 0.20, 0.30, 0.40, and 0.50, respectively.

From the covertype we extracted six datasets, with a number of instances ranging from 39,000 to 237,000 and a size ranging from 5 MB to 30 MB. For each of those six datasets as input, we executed the same workflow with Weka and Weka4WS. The machine used for this experiment has

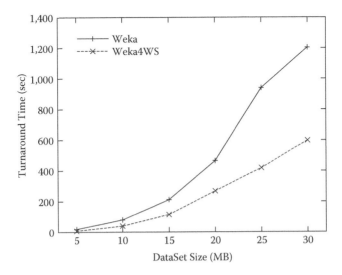

FIGURE 8.9 Turnaround times of a classification workflow on a two-processor dual-core machine: Weka4WS versus Weka.

two Intel Xeon dual-core processors with a clock frequency of 3 GHz and 2 GB of RAM. The turnaround times are reported in Figure 8.9.

On Weka, the turnaround time ranges from 19 seconds for the smallest data, to about 20 minutes for the largest dataset. Using Weka4WS the turnaround time increases from 8 seconds to about 10 minutes, thus saving an amount of time ranging from a minimum of 42% to a maximum of 58%. These results show that the multithreaded approach of Weka4WS is well suited to exploit the computational power of multiprocessor and multicore machines.

8.3 WEB SERVICES RESOURCE FRAMEWORK (WSRF) OVERHEAD IN DISTRIBUTED SCENARIOS

The main goal of this section is the evaluation of the overhead introduced by the WSRF mechanisms with respect to the total execution time, with the aim of assessing the efficiency of the service-oriented approach in distributed knowledge discovery scenarios. For the sake of explanation we refer here to the Weka4WS system, but the results discussed in the following are also valid for the Knowledge Grid services, because they are based on invocation mechanisms similar to those adopted by the Weka4WS services.

To carry out the evaluation, we implemented a simple distributed application in which a Weka4WS client asks a remote Weka4WS service to perform a clustering analysis on a dataset located on the user node (Talia, Trunfio, and Verta, 2008). This is similar to the scenario described in Section 6.5, where we analyzed all the steps (*resource creation, notification subscription, task submission, file transfer, data mining, notification reception, results retrieving,* and *resource destruction*) performed to execute a single data mining task in the Weka4WS framework. Our goal here is measuring the amount of time needed to perform each of these steps, so as to evaluate how they affect the total execution time.

As input data source we used the USCensus1990 dataset,* which contains data extracted from the U.S. Census Bureau Web site as part of the 1990 U.S. census. From the original data source we extracted 10 datasets containing a number of instances ranging from 143,000 to 1,430,000, with a size ranging from 20 MB to 200 MB. Then, we used Weka4WS to remotely execute the Expectation Maximization (EM) clustering algorithm on each of these datasets, so as to group data into five clusters on the basis of 10 selected attributes.

The application has been executed in two Grid scenarios:

- *Local Area Grid (LAG):* The computing node and the user node are connected by a local area network with an average bandwidth of 41.2 MB/s.

- *Wide Area Grid (WAG):* The computing node and the user node are connected by a wide area network with an average bandwidth of 52.7 kB/s.

For each dataset size and for each Grid scenario, we run 20 independent executions. The measures reported in Figure 8.10 through Figure 8.13 resulted from the average values of all the executions. Figure 8.10 shows the execution times of each step in the LAG scenario for the different dataset sizes. For measurement purposes, the *notification reception* and *results retrieving* steps are hereafter cumulatively referred to as *results delivery.*

As expected, the execution times of the WSRF-specific steps are independent from the dataset size, namely *resource creation* (1.7 seconds, on average), *notification subscription* (0.26 seconds), *task submission* (0.29 seconds), *results delivery* (1.3 seconds), and *resource destruction* (0.21 seconds). On the contrary, the execution times of the *file transfer*

---

* UCI KDD Archive, 1990 U.S. Census Data, http://kdd.ics.uci.edu/databases/census1990/USCensus1990.html (accessed June 2012).

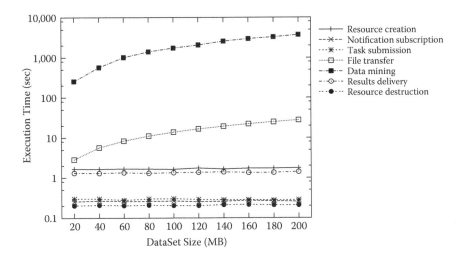

FIGURE 8.10   Execution times of the different steps in the Local Area Grid (LAG) scenario.

and *data mining* steps are proportional to the dataset size. In particular, the execution time of the *file transfer* ranges from 2.8 seconds for 20 MB to 28 seconds for 200 MB, while the *data mining* execution time ranges from 4.2 minutes for the dataset of 20 MB, to 61.8 minutes for the dataset of 200 MB. The total execution time (sum of the execution times of all the steps) ranges from 4.4 minutes for 20MB, to 62.3 minutes for 200 MB.

Figure 8.11 shows the execution times of all the steps in the WAG scenario. The execution times of the WSRF-specific steps are similar to those measured in the LAG scenario. The main difference is the execution time of the *results delivery* step, which in the LAG scenario is on average 1.3 seconds while in the WAG scenario is equal to 2.7 seconds, due to additional time needed to transfer the clustering model through a wide area network. For the same reason, the transfer of the dataset to be mined requires an execution time significantly greater than the one measured in the LAG scenario. In particular, the execution time of the *file transfer* step in the WAG scenario passes from about 1 minute for 20 MB, to 10.5 minutes for 200 MB.

The *data mining* execution time is similar to that measured in the LAG scenario, because the clustering analysis is executed on an identical computing node. Mainly due to the additional time required by the *file transfer* step, the total execution time is greater than the one measured in the LAG scenario; it ranges from 5.3 minutes for the 20 MB dataset to 70.7 minutes for the 200 MB dataset.

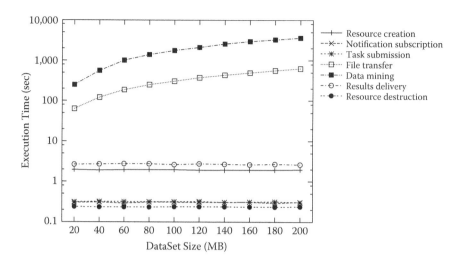

FIGURE 8.11   Execution times of the different steps in the Wide Area Grid (WAG) scenario.

To better highlight the overhead introduced by the WSRF mechanisms and the distributed scenario, Figures 8.12 and 8.13 show the percentage of *data mining*, *file transfer,* and *WSRF overhead* (i.e., the sum of *resource creation, notification subscription, task submission, results delivery,* and *resource destruction* steps) with respect to the total execution time in the LAG and WAG scenarios.

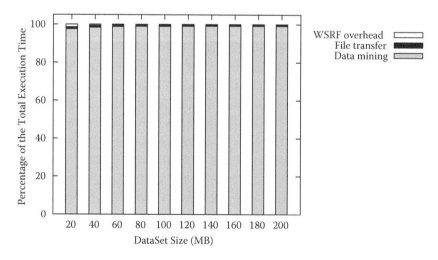

FIGURE 8.12   Percentages of the total execution time of the different steps in the Local Area Grid (LAG) scenario.

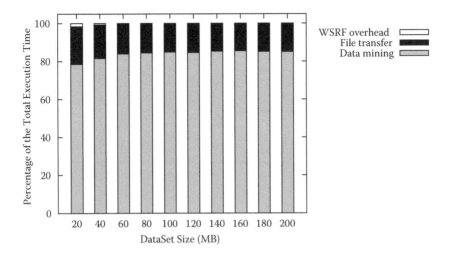

FIGURE 8.13   Percentages of the total execution time of the different steps in the Wide Area Grid (WAG) scenario.

In the LAG scenario, the *data mining* step takes 97.5% of the total execution time for the dataset of 20 MB, while it takes 99.2% of the total execution time for the dataset of 200 MB. At the same time, the *file transfer* ranges from 1.1% to 0.7%, and the *WSRF overhead* ranges from 1.4% to 0.1%. In the WAG scenario, the *data mining* step takes from 78.5% to 85.1% of the total execution time, the *file transfer* ranges from 19.8% to 14.8%, while the *WSRF overhead* ranges from 1.7% to 0.1%.

We can observe that in the LAG scenario, neither the *WSRF overhead* nor the *file transfer* represent a significant part of the total execution time. In the WAG scenario, the *WSRF overhead* is a negligible factor, while the *file transfer* can be a critical step, particularly with small datasets for which the *data mining* step is fast. However, in most scenarios the data mining step is a very time-consuming task; hence, in these cases, the *file transfer* step, if needed, becomes marginal compared to the data mining step.

As a concluding remark, the experimental results discussed above show that the WSRF overhead is negligible in all practical scenarios, thus demonstrating the efficiency of this service-oriented technology for data analysis applications in distributed environments.

CHAPTER 9

Sketching the Future Pervasive Data Services

THIS CHAPTER OUTLINES AND DISCUSSES the role of service-oriented infrastructures in providing future ubiquitous services for data access, distribution, processing, and analysis. The requirements, the features, and the importance of service-oriented decentralized data-oriented techniques in ubiquitous computing, smart environments, and ambient intelligence applications are described. The main goal of this chapter is to envision new opportunities and application scenarios for data management and knowledge discovery by using new-generation service-oriented platforms that are mobile, pervasive, large scale, and can be composed of a massive number of devices according to the future Internet paradigm.

9.1 SERVICE ORIENTATION AND UBIQUITOUS COMPUTING FOR DATA

Service-oriented infrastructures like Grids and Clouds extend the network paradigm offered by the Internet and offer a novel way to think about and use computing resources and services. In particular, as data sources available across the Internet are growing at a tremendous pace, new distributed, dynamic and scalable solutions for data management and analysis are needed to unlock the value of data and power of knowledge that are hidden in it.

Today the term *Big data* is used to refer to large-scale data volumes that are exponentially growing and are available through the Internet. This enormous amount of data encourages development of scalable techniques, models, services, and applications that must be run on distributed high-performance infrastructures and provide pervasive access to data and analysis tools for making discoveries in science and for intelligent business applications that will produce economic value from available data.

Grids and Clouds are distributed, heterogeneous, and dynamic. They can be effectively used for scalable computing and data storage. Therefore, they can be effectively integrated with computing models such as ubiquitous systems and ambient intelligence. Smart territories envision a future scenario where humans will be surrounded by sensitive and responsive computing environments. Ambient intelligence, pervasive computing, and smart cities technologies combine concepts of ubiquitous computing and intelligent systems serving people in their daily lives. In doing this, ambient intelligence represents an advancement of concepts and environments that have been developed, such as mobile systems, ubiquitous computing, service-oriented computing, and ambient computing.

The concept of ambient computing implies that the computing environment is always present and available in an even manner. The concept of ubiquitous computing implies that the computing environment is available everywhere and is "into everything." The concept of mobile computing implies that the end-user device may be connected even when on the move. Ambient intelligence integrates those concepts into intelligent systems solutions that provide learning algorithms and pattern matchers, metadata and semantic models, speech recognition and language translators, and gesture classification and situation assessment.

Computational Grids enable the sharing, selection, and aggregation of a wide variety of geographically distributed computing and data resources and present them as a single, unified resource for solving large-scale compute- and data-intensive applications. Due to its nature, a Grid can offer effective support for applications running on geographically dispersed computing devices providing services for user authentication, security, data privacy, and decentralized cooperation. Although Grids include computing resources connected by wired communication media, in the future wireless Grids will be implemented and deployed. This is the first step toward future ubiquitous service-oriented computing and ambient intelligence Grids and Clouds.

On the one hand, Grids and Clouds might represent the backbone for the implementation of compute- and data-intensive ambient intelligence applications that cover large geographical areas of local settings where intelligent services must be provided to people. On the other hand, they might represent the computing paradigm that can be used in self-organizing and self-configuring ambient intelligence systems that can be arranged through the coordination and cooperation of dispersed computing resources such as personal digital assistants (PDAs), sensors, mobile devices, radio-frequency identification (RFID) tags, data repositories, intelligent interfaces, and so forth. We can envision a possible evolution of service-oriented infrastructures from today's Grids, Clouds, and Web servers, which are essentially wired networks of large-grained computers, to a long-term scenario of an ambient intelligence Internet that will integrate wireless and ubiquitous infrastructures to provide knowledge-based services in a pervasive way. In particular, we can define the basic features of four Grid and Cloud scenarios:

- *Wired Grids/Clouds*: Current off-the-shelf Grids/Clouds coordinating large-grained computing elements connected through wired networks

- *Wireless Grids and mobile-to-Cloud systems*: Grids that are able to dynamically aggregate computing devices connected through wireless networks by combining mobile technology with wired Grid and Cloud middleware

- *Ubiquitous Grids/Clouds*: Next-generation infrastructures featuring high mobility and high embeddedness where sensor networks, mobile devices, and so forth, can be dynamically connected to extend their original scope

- *Ambient Intelligence Grids and InterClouds*: Future Grids and InterClouds able to support requirements of ubiquitous systems leveraging ambient intelligence techniques (e.g., intuitive interfaces, context-awareness, knowledge discovery, and ambient semantics) and novel service-oriented middleware

9.2 TOWARD FUTURE SERVICE-ORIENTED INFRASTRUCTURES

Large-scale distributed computing infrastructures have often evolved through different technological phases that were not originally planned. This trend occurred to the Web and later to Grids and Clouds. For example, since their birth, computational Grids have traversed different phases or generations (De Roure et al., 2003; Cannataro, Talia, and Trunfio, 2006). In the early 1990s, *first-generation Grids* enabled the interconnection of large supercomputing centers to obtain an aggregate computational power greater than that offered by participating sites, or to execute distributed computations over thousands of workstations introducing the first real implementation of a metacomputing model. *Second-generation Grids* were characterized by the adoption of standards (such as Hypertext Transfer Protocol [HTTP], Lightweight Directory Access Protocol [LDAP], Public Key Infrastructure [PKI], etc.) that enabled the deployment of a global-scale computing infrastructure linking remote resources. Second-generation Grids started to use a Grid middleware as glue between heterogeneous distributed systems, resources, users, and local policies. Grid middleware targeted technical challenges such as communication, scheduling, security, information, data access, and fault detection. Main representatives of second-generation Grids were systems like Globus Toolkit, Unicore, and Condor. A milestone between second- and *third-generation Grids* was posed by Foster, Kesselman, and Tuecke, who defined the Grid as "flexible, secure, coordinated resource sharing among dynamic collections of individuals, institutions, and resources— what we refer to as virtual organizations" (2001, 200). The motivation for third-generation Grids was to simplify and structure the systematic building of Grid applications through the composition and reuse of software components and the development of knowledge-based services and tools. Therefore, following the trend that has emerged in the Web community, the service-oriented model has been proposed.

A major factor that will further drive the evolution of Grids and Clouds is the necessity to face the enormous amount of data that any field of human activity is producing at a rate never seen before. The obstacle is not the technology to store and access data, but perhaps what is lacking is the ability to transform data stores into useful information, extracting knowledge from them (Fayyad and Uthurusamy, 2002).

In our vision, the architecture of *future generation service-oriented infrastructures*, will be based on the convergence of technologies, methodologies, and solutions that are emerging in many computer science fields, apparently far away from and unaware of Grids and Clouds, such as mobile and pervasive computing, agent-based systems, ontology-based reasoning, peer-to-peer, data analytics, and knowledge management. An example of an "infrastructural" methodology that is becoming common practice in many of the previous fields is the systematic adoption of metadata to describe resources, services, and data sources, and to enhance, and possibly automate, strategic processes such as service discovery and negotiation, application composition, information extraction, and knowledge discovery.

Although the ongoing convergence between Grids, Clouds, Web services, and the Semantic Web represents a milestone toward a worldwide service-oriented architecture, which has the potential to face important issues such as Big data handling, business modeling, and application programming, many other issues need research and development efforts. To be effectively adopted in different application domains, future generation infrastructures need to address different issues such as an increasing complexity and distribution of applications; different goals, skills, and habits of users; availability of different programming and deployment models; heterogeneous capabilities; and performances of access networks and devices.

More specifically, the great availability of digital data and information; the maturity of data exploration techniques able to extract and synthesize knowledge, such as data mining, text summarization, semantic modeling, and knowledge management; and the demand for intelligent services in different phases of the application life cycle are the driving forces toward novel ubiquitous knowledge-based services.

Scientific and commercial applications, as well as middleware, will increasingly produce an overwhelming quantity of application and usage data that need to be exploited. The way such data, at different levels, can be effectively acquired, represented, exchanged, integrated, and converted into useful knowledge is an emerging research field. In particular, ontologies (Gruber, 1993) and metadata (Keenoy, Poulovassilis, and Christophides, 2003) are the basic elements through which smart services can be developed. Using ontologies, Web servers, Clouds, and Grids may offer semantic modeling of user's tasks/needs, available services, and data sources. Semantic modeling of resources, services, and data is the enabling factor to support some important emerging services, such as dynamic service discovery (Cannataro and Comito, 2003), composition, and scheduling, which can be used to enhance tools for workflow composition and enactment.

Moreover, data mining and knowledge management techniques will enable intelligent services based on the exploration and summarization of stored data. Such services could be employed at both the operation layer, where infrastructure management could gain from information hidden in usage data, and the application layer, where users could be able to exploit distributed data repositories, using the distributed computing platforms not only for high-performance data access, transfer, and processing, but also to apply key analysis tools and instruments. Examples of knowledge discovery and knowledge management services are Grid-aware information systems (e.g., document management and knowledge extraction on a Grid), and Cloud knowledge-based middleware services (e.g., scheduling and resource allocation based on resources usage patterns). Finally, context-awareness and autonomous behavior, often required in ambient intelligence applications, can be implemented by combining novel analysis of infrastructure usage data with well-established computational and interaction models such as those offered by intelligent agents (Foster, Jennings, and Kesselman, 2004).

9.3 REQUIREMENTS OF FUTURE GENERATION SERVICES

In summary, given the current and future challenges, the main requirements of future generation services are:

- *Semantic modeling applied extensively to each component of the infrastructure,* including services, applications, data sources, and computing devices (from ambient sensors to high-performance computers).

- *Modeling of user's tasks/needs* to enable adaptation and personalization not only in content delivery, but also in performance and quality offered to services and applications.

- *Advanced forms of collaboration,* through dynamic formation and negotiation of virtual organizations and agent technologies.

- *Self-configuration,* autonomic management, dynamic resource discovery, and fault tolerance (e.g., to support the seamless integration of emerging wireless infrastructures).

In a service-oriented architecture, some novel knowledge-based services that will take care of those requirements are:

- *Knowledge discovery and knowledge management services,* for both user's needs (e.g., intelligent exploration of data) and system management (e.g., intelligent use of resources).

- High-level services for dynamic service discovery, composition, and scheduling.

- Services for supporting pervasive and ubiquitous computing through environment/context awareness and adaptation.

- *Autonomous behavior*, especially when there is lack of complete information (e.g., about environment, user goals, etc.).

9.4 SERVICES FOR UBIQUITOUS COMPUTING

There is a twofold relationship between service-oriented architecture (SOA) platforms and ubiquitous computing. On the one hand, Web, Grid, and Cloud applications often require ubiquitous access to geographically distributed resources, including computing elements, data repositories, sensor networks, and mobile devices. We can envision such *ubiquitous platforms* as composed of a relatively static network of large-grain nodes, to which sensor networks, mobile devices, and so forth, can be dynamically connected to extend the scope of the platform.

From another point of view, large-scale ubiquitous applications may involve typical issues such as authentication, authorization, cryptography, service description, resource discovery, and resource management, and can sometimes require the computational power of Grids and Clouds to accomplish their goals. Grid and Cloud technologies and models can thus provide the needed computational infrastructure and services to address these challenges to build ubiquitous environments.

In analyzing the first scenario, the main issue to be faced is the dynamic joining of devices to a wired infrastructure. In current systems, it is often assumed that resources are usually statically connected to the network with well-known Internet Protocol (IP) addresses, reliable network connections, and have enough computational power to host "heavy-weight" middleware. Moreover, current systems are designed to build organization-based environments, where a set of "enrollment" procedures have to be accomplished when a node joins a system. Furthermore, system functions are often based on hierarchical models and services (e.g., resource publishing and discovery) that assume trust relationships among resources.

In ubiquitous service-oriented systems it may not be feasible to guarantee trust about resources, because devices, in general, cannot be univocally identified, and they could appear and disappear in different places and environments (e.g., an iPad moving across organizations). Moreover, as stated before, current middleware is usually huge so it is not well suited to run on most of the small devices typically involved in ubiquitous systems. Similar to operating systems and software platforms that are often scaled down to run on small devices (e.g., Windows Mobile, Embedded Linux, and J2ME), middleware and inner procedures should be rethought to be applicable to such devices. Moreover, strict security and interaction models should be adapted to enable intermittent participation of trusted and untrusted devices.

When considering the second perspective, the main issue to be faced is defining appropriate models and approaches to the dynamic, on-demand, self-configuring aggregation of groups of devices able to act as a distributed platform to accomplish possibly complex tasks. A scenario of interest is a user (i.e., his or her PDA) who demands for a peak of computational power or a large amount of storage for his or her data, which can be satisfied by aggregating the power of devices available in the surrounding environment. Other than using vertically integrated solutions, which are not scalable and reusable, a possible solution is joining a ubiquitous platform for accessing its services and exploiting the available computing power.

Some feasible approaches to enable ubiquitous infrastructures are:

- Provision of *access nodes* to support dynamic joining and leaving of intermittent devices. Such brokers should be equipped in such a way that they can support protocols and networks of ubiquitous systems, and update the system state.

- Provision of *small-footprint middleware* to be run, possibly in a dynamic on-demand way, on devices of ubiquitous systems. Such middleware should take into account the features of ubiquitous systems described so far, and should be based on nonhierarchical decentralized protocols, technologies supporting code movement, such as agents, and of course on small-footprint operating systems.

- Definition of an *ubiquitous platform model* through which devices of an ubiquitous infrastructure can behave as a service-oriented network (e.g., taking into consideration well-established mobile communications and computing models).

In summary, the relationships between service-oriented systems and ubiquitous computing systems can be exploited in two directions. According to the first direction, the service-oriented infrastructures represent a backbone for ubiquitous computing services and applications that provide high performance and large data storage and communication facilities; in line with the second one, a self-organization of ubiquitous computing devices can be the basis for composing dynamic ubiquitous services by exploiting the infrastructure.

9.5 SERVICES FOR AMBIENT INTELLIGENCE AND SMART TERRITORIES

As discussed, ambient intelligence achievement needs the exploitation of advanced distributed computing systems and technologies that will support the development and delivery of intelligent services in geographically distributed environments where humans operate. Wired infrastructures, together with wireless and ubiquitous platforms, can be integrated to provide the distributed computing infrastructure for delivering intelligent services that are invisible to final users.

Service-oriented ambient intelligence implemented by Grids, Clouds, mobile systems, and smart objects involves a seamless environment of computing, advanced communication technology, semantic-rich description of resources and services, and intelligent interfaces (Jeffery, 2003). Because an ambient intelligent system is aware of the particular characteristics of human presence and roles, it must take care of needs, is capable of responding intelligently to human requests, and can even engage in intelligent dialogue. This process requires the use of artificial intelligence and knowledge discovery techniques and systems that often need efficient computing resources and high-bandwidth communications to support advanced intelligent services. Today's service-oriented models and future network infrastructures are the best candidates to provide ubiquitous support to run complex software systems and efficient communication protocols, as required for the implementation of ambient intelligence systems.

Ambient intelligence should also be unobtrusive and most often invisible. Users should use it where needed without seeing its infrastructure all the time. Moreover, interaction with ambient intelligent systems should be comforting and agreeable for humans. These properties require the availability of high-level hardware and software interfaces that represent the front-end of intelligent applications. Those interfaces also require high performance and large data management functionality that can be provided by service-oriented systems like Grids and Clouds.

Service-oriented ambient intelligence requires the integrated use of several advanced technologies for supporting high-performance decentralized computing and for providing pervasive data and knowledge access and management. Here we list some of those key research areas that will provide solutions that can be part of ambient intelligence systems:

- *Intelligent agents*: Agents are autonomous problem solvers that can act flexibly in uncertain and dynamic environments; there is a convergence of interests, with agent systems requiring robust infrastructure, and high-performance computing systems requiring autonomous, flexible behaviors. Embodying intelligence in multiagent systems requires a high-performance support that can be offered by high-performance computing systems, Grids, and Clouds (Talia, 2012); at the same time, intelligent agents represent a key model for ambient intelligence systems implementations (Foster, Jennings, and Kesselman, 2004).

- *Decentralized computing*: When a large number of dispersed nodes must cooperate, decentralized models and peer-to-peer networks can be used to implement scalable systems without centralized points; potential convergence between SOA systems and peer-to-peer models can provide scalable architectural solutions.

- *Self-organizing systems*: Applications composed of a large number of computing elements distributed in a geographic area can benefit from self-organization features that, according to emergent behavior computing, build the system functionality by integrating the single node operations.

- *Sensor networks*: Sensors are devices largely used in ambient computing frameworks to receive inputs from the environment and to register its state; sensor networks are part of the network infrastructure supporting ambient intelligence and smart territories. Service-oriented systems can be effectively used to provide data storage and a computing backbone for sensor networks.

- *Metadata and ontologies for ambient intelligence*: Decentralized systems involve multiple entities that cooperate to ask for and to deliver services. Meaningful interactions are difficult to achieve in any open system because different players typically have distinct information models. Advances are required in areas such as metadata definition, common ontology definition, schema mediation, and semantic mediation for ambient intelligence applications.

All the technologies listed above are investigated both in ambient intelligence and in service-oriented computing. Therefore, advancements in those areas will be embodied in future computing systems. This process can be a further step toward Grid- and Cloud-based services for ubiquitous knowledge discovery and ambient intelligence.

9.6 CONCLUSIVE REMARKS

New service-oriented computing platforms like Grids and Clouds are key components of a new Internet computing infrastructure that allows users to access remote computing, data resources, and facilities that are not available in a single site. By linking users, computers, databases, sensors, and other devices, these distributed systems extend the network paradigm offered by the Internet and the Web, and offer a novel way to think and use computing. Service-oriented networks are distributed, heterogeneous, and dynamic, and so they can be effectively integrated with similar computing models such as ubiquitous systems and ambient computing. Service-oriented Grids and InterClouds can represent the backbone for implementing ubiquitous computing services. Therefore, if ambient intelligence systems are implemented on top of such service-oriented infrastructures, end users can access anytime, anyhow, anywhere computation resources, data, and knowledge to implement advanced applications and to use future information technology infrastructures and environments at the Internet scale.

Bibliography

Adaçal, M., and A. B. Bener. 2006. Mobile Web services: A new agent-based framework. *IEEE Internet Computing* 10(3):58–65.

Agrawal, G. 2003. High-level interfaces and abstractions for Grid-based data mining. *Workshop on Data Mining and Exploration Middleware for Distributed and Grid Computing*, Minneapolis, MN.

Agrawal, R., T. Imielinski, and A. Swami. 1993. Mining association rules between sets of items in large databases. In *Proceedings of the ACM SIGMOD International Conference on Management of Data*, ed. P. Buneman and S. Jajodia, 207–216, Washington, DC. New York: ACM Press.

Agrawal, R., and J. C. Shafer. 1996. Parallel mining of association rules. *IEEE Transactions on Knowledge and Data Engineering* 8(6):962–969.

Agrawal, R., and R. Srikant. 1994. Fast algorithms for mining association rules. In *Proceedings of the 20th International Conference on Very Large Databases*, ed. J. B. Bocca, M. Jarke, and C. Zaniolo, 487–499, Santiago Chile. San Francisco: Morgan Kaufmann.

Alcamo, P., F. Domenichini, and F. Turini. 2000. An XML-based environment in support of the overall KDD process. In *Proceedings of the 4th International Conference on Flexible Query Answering Systems*, ed. H. L. Larsen, J. Kacprzyk, S. Zadrozny, T. Andreasen, and H. Christiansen, 413–424, Warsaw, Poland. Heidelberg, Germany: Physica-Verlag.

Allcock, W., J. Bresnahan, R. Kettimuthu, M. Link, C. Dumitrescu, I. Raicu, and I. Foster. 2005. The Globus striped GridFTP framework and server. *Proceedings of the ACM/IEEE SC2005 Conference on High Performance Networking and Computing*, 54, Seattle, WA. Los Alamitos, CA: IEEE Computer Society.

Altintas, I., C. Berkley, E. Jaeger, M. Jones, B. Ludascher, and S. Mock. 2004. Kepler: An extensible system for design and execution of scientific workflows. *Proceedings of the 16th International Conference on Scientific and Statistical Database Management*, 423–424, Santorini Island, Greece.

Berman, F. 2001. From TeraGrid to Knowledge Grid. *Communications of the ACM* 44(11):27–28.

Beynon, M., T. Kurc, U. Catalyurek, C. Chang, A. Sussman, and J. Saltz. 2001. Distributed processing of very large datasets with DataCutter. *Parallel Computing* 27(11):1457–1478.

Bigus, J. P. 1996. *Data Mining with Neural Networks*. New York: McGraw-Hill.

Birant, D. 2011. Service-oriented data mining. In *New Fundamental Technologies in Data Mining*, ed. K. Funatsu, InTech. http://www.intechopen.com/articles/show/title/service-oriented-data-mining (accessed June 2012).

Brezany, P., I. Janciak, and A. Min Tjoa. 2008. GridMiner: An Advanced Support for e-Science Analytics. In *Data Mining Techniques in Grid Computing Environments*, ed. W. Dubitzky, 37–55. New York: John Wiley & Sons.

Bruynooghe, M. 1989. Parallel implementation of fast clustering algorithms. In *Proceedings of the International Symposium on High Performance Computing*, ed. J. L. Delhaye and E. Gelenbe, 65–78, Montpellier, France. Amsterdam: North-Holland.

Bueti, G., A. Congiusta, and D. Talia. 2004. Developing distributed data mining applications in the Knowledge grid framework. In *Proceedings of the 6th International Conference on High Performance Computing for Computational Science*, ed. M. J. Daydé, J. Dongarra, V. Hernández, J. M. Laginha, and M. Palma, 156–169, Valencia, Spain. Berlin, Germany: Springer.

Cannataro, M., and C. Comito. 2003. A data mining ontology for Grid programming. *Proceedings of the 1st International Workshop on Semantics in Peer-to-Peer and Grid Computing*, 113–114, Budapest, Hungary.

Cannataro, M., C. Comito, A. Congiusta, and P. Veltri. 2004. PROTEUS: A bioinformatics problem-solving environment on Grids. *Parallel Processing Letters* 14(2):217–237.

Cannataro, M., and D. Talia. 2003. The Knowledge Grid. *Communications of the ACM* 46(1):89–93.

Cannataro, M., and D. Talia. 2004. Semantics and knowledge grids: Building the next-generation grid. *IEEE Intelligent Systems* 19(1):56–63.

Cannataro, M., D. Talia, and P. Trunfio. 2001. KNOWLEDGE GRID: High-performance knowledge discovery services on the Grid. In *Proceedings of the 2nd International Workshop on Grid Computing*, ed. Ca. A. Lee, 38–50, Denver, CO. Berlin, Germany: Springer.

Cannataro, M., D. Talia, and P. Trunfio. 2006. Grids for Ubiquitous Computing and Ambient Intelligence. In *Ambient Intelligence, Wireless Networking, Ubiquitous Computing*, ed. A. Vasilakos and W. Pedrycz, 127–142. Norwood, MA: Artech House.

Catlett, C. 2002. The TeraGrid: A Primer. http://teragridforum.org/mediawiki/images/b/b5/TeraGrid_Primer_September_2002.pdf (accessed June 2012).

Cesario, E., M. Lackovic, D. Talia, and P. Trunfio. 2011. Service-Oriented Data Analysis in Distributed Computing Systems. In *High Performance Computing: From Grids and Clouds to Exascale*, ed. I. Foster, W. Gentzsch, L. Grandinetti, and G. Joubert, 225–245. Lansdale, PA: IOS Press.

Cesario, E., M. Lackovic, D. Talia, and P. Trunfio. 2012. Programming Knowledge Discovery Workflows in Service-Oriented Distributed Systems. Submitted for publication.

Chapman, P., J. Clinton, R. Kerber et al. 2000. CRISP-DM 1.0 step-by-step data mining guide. Technical report. The CRISP-DM consortium.

Charniak, E., and D. McDermott. 1985. *Introduction to Artificial Intelligence.* Reading, MA: Addison-Wesley.

Cheeseman, P., and J. Stutz. 1996. Bayesian Classification (AutoClass): Theory and Results. In *Advances in Knowledge Discovery and Data Mining*, ed. U. M. Fayyad, G. Piatetsky-Shapiro, P. Smyth, and R. Uthurusamy, 61–83. Cambridge, MA: AAAI Press/MIT Press.

Chen, M.-S., J. Han, and P. S. Yu. 1996. Data mining: An overview from a database perspective. *IEEE Transactions on Knowledge and Data Engineering* 8(6):866–883.

Chu, H., C. You, and C. Teng. 2004. Challenges: Wireless Web services. *Proceedings of the 10th International Conference on Parallel and Distributed Systems (ICPADS 2004)*, 657–664, Newport Beach, CA. Los Alamitos, CA: IEEE Computer Society.

Comito, C., D. Talia, and P. Trunfio. 2010. A distributed architecture for energy-efficient data mining over mobile devices. In *Proceedings of the COST Action IC0804 on Large Scale Distributed Systems—First Year*, J-M. Pierson and H. Hlavacs, 28–31.

Comito, C., D. Talia, and P. Trunfio. 2011. An energy-aware clustering scheme for mobile applications. *Proceedings of the 11th IEEE International Conference on Scalable Computing and Communications*, 15–22, Paphos, Cyprus. Los Alamitos, CA: IEEE Computer Society.

Congiusta, A., D. Talia, and P. Trunfio. 2003. VEGA: A visual environment for developing complex Grid applications. In *Proceedings of the 1st International Workshop on Knowledge Grid and Grid Intelligence*, ed. W. K. Cheung and Y. Ye, 56–66, Halifax, Canada. Halifax, Canada: Saint Mary's University.

Congiusta, A, D. Talia, and P. Trunfio. 2006. A Visual Programming Environment for Developing Complex Grid Applications. In *Grid Computing: Software Environments and Tools*, ed. J. Cunha and O. Rana, 257–283. New York: Springer.

Congiusta, A., D. Talia, and P. Trunfio. 2007. Distributed data mining services leveraging WSRF. *Future Generation Computer Systems* 23(1):34–41.

Congiusta, A., D. Talia, and P. Trunfio. 2008. Service-oriented middleware for distributed data mining on the Grid. *Journal of Parallel and Distributed Computing* 68(1):3–15.

Curcin, V., M. Ghanem, Y. Guo et al. 2002. Discovery Net: Toward a Grid of knowledge discovery. *Proceedings of the 8th International Conference on Knowledge Discovery and Data Mining*, 658–663, Edmonton, Canada. New York: ACM Press.

Czajkowski, K., D. F. Ferguson, I. Foster, J. Frey, S. Graham, I. Sedukhin, D. Snelling, S. Tuecke, and W. Vambenepe. 2004. The WS-Resource Framework. http://www.globus.org/wsrf/specs/ws-wsrf.pdf (accessed June 2012).

Deelman, E., J. Blythe, Y. Gil, C. Kesselman, G. Mehta, S. Patil, M.-H Su, K. Vahi, and M. Livny. 2004. Pegasus: Mapping scientific workflows onto the Grid. In *Proceedings of the 2nd European Across Grids Conference*, ed. M. D. Dikaiakos, 11–20, Nicosia, Cyprus. Berlin, Germany: Springer.

De Roure, D., M. A. Baker, N. R. Jennings, and N. R. Shadbolt. 2003. The Evolution of the Grid. In *Grid Computing: Making the Global Infrastructure a Reality*, ed. F. Berman, G. Fox, and A. Hey, 65–100. New York: Wiley.

Dimopoulos, Y., and A. Kakas. 1996. Abduction and Learning. In *Advances in Inductive Logic Programming*, ed. L. De Raedt, 144–171. Berlin: IOS Press.

Dumas, M., and A. H. M. ter Hofstede. 2001. UML activity diagrams as a work-flow specification language. In *Proceedings of UML 2001*, M. Gogolla and C. Kobryn, 76–90, Toronto, Canada. Berin, Germany: Springer.

Enright, A. J., S. V. Dongen, and C. A. Ouzounis. 2002. An efficient algo-rithm for large-scale detection of protein families. *Nucleic Acids Research* 30(7):1575–1584.

Fahlman, S. E., and G. E. Hinton. 1987. Connectionist architectures for artificial intelligence. *IEEE Computer* 20(1):100–109.

Fahringer, T., A. Jugravu, S. Pllana, R. Prodan, C. Seragiotto Junior, and H. L. Truong. 2005. ASKALON: A tool set for cluster and Grid computing. *Concurrency and Computation: Practice & Experience* 17(2–4):143–169.

Fayyad, U. M., G. Piatetsky-Shapiro, and P. Smyth. 1996. From Data Mining to Knowledge Discovery: An Overview. In *Advances in Knowledge Discovery and Data Mining*, ed. U. M. Fayyad, G. Piatetsky-Shapiro, P. Smyth, and R. Uthurusamy, 1–34. Menlo Park, CA: AAAI/MIT Press.

Fayyad, U. M., and R. Uthurusamy. 1996. Data mining and knowledge discovery in databases. *Communications of the ACM* 39(11):24–26.

Fayyad, U. M., and R. Uthurusamy (ed.). 2002. Evolving data mining into solu-tions for insights (Special issue on). *Communications of the ACM* 45(8).

Fogelman Soulie, F. 1991. Neural networks and computing. *Future Generation Computer Systems* 7(1):69–77.

Foster, I. 2005. Globus Toolkit Version 4: Software for service-oriented sys-tems. In *Proceedings of the 2nd International Conference on Network and Parallel Computing*, ed. H. Jin, D. Reed, and W. Jiang.2–13, Beijing, China. Berlin, Germany: Springer.

Foster, I., N. R. Jennings, and C. Kesselman. 2004. Brain meets brawn: Why Grid and agents need each other. *Proceedings of the 3rd International Joint Conference on Autonomous Agents and Multiagent Systems*, 8–15, New York. Los Alamitos, CA: IEEE Computer Society.

Foster, I., and C. Kesselman. 1998. Computational Grids. In *The Grid: Blueprint for a New Computing Infrastructure*, ed. I. Foster and C. Kesselman, 15–52. Los Altos, CA: Morgan Kaufmann.

Foster, I., C. Kesselman, J. M. Nick, and S. Tuecke. 2002. Grid services for distrib-uted system integration. *IEEE Computer* 35(6):37–46.

Foster, I., C. Kesselman, J. M. Nick, and S. Tuecke. 2003. The Physiology of the Grid. In *Grid Computing: Making the Global Infrastructure a Reality*, ed. F. Berman, G. Fox, and A. Hey, 217–249. New York: Wiley.

Foster, I., C. Kesselman, and S. Tuecke. 2001. The anatomy of the Grid: Enabling scalable virtual organizations. *International Journal of Supercomputing Applications* 15(3):200–222.

Foti, D., D. Lipari, C. Pizzuti, and D. Talia. 2000. Scalable parallel clustering for data mining on multicomputers. In *Proceedings of the 3rd International Workshop on High Performance Data Mining*, ed. J. D. P. Rolirn, 390–398, Cancun, Mexico. Berlin, Germany: Springer.

Giannadakis, N., A. Rowe, M. Ghanem, and Y. Guo. 2003. InfoGrid: Providing information integration for knowledge discovery. *Information Sciences* 155:199–226.

Grossman, R. L., Y. Gu, C. Gupta, D. Hanley, X. Hong, and P. Krishnaswamy. 2004. Open DMIX: High performance Web services for distributed data mining. *7th International Workshop on High Performance and Distributed Mining*, Orlando, FL.

Gruber, T. R. 1993. A translation approach to portable ontologies. *Knowledge Acquisition* 5(2):199–220.

Guedes, D., W. Meira, and R. Ferreira. 2006. Anteater: A service-oriented architecture for high-performance data mining. *IEEE Internet Computing* 10(4):36–43.

Hall, M., E. Frank, G. Holmes, B. Pfahringer, P. Reutemann, and I. H. Witten. 2009. The WEKA Data Mining Software: An Update. *SIGKDD Explorations* 11(1):10–18.

Han, E. H., G. Karypis, and V. Kumar. 2000. Scalable parallel data mining for association rules. *IEEE Transactions on Knowledge and Data Engineering* 12(2):337–352.

Hinke, T., and J. Novonty. 2000. Data mining on NASA's Information Power Grid. *Proceedings of the 9th International Symposium on High Performance Distributed Computing*, 292–293, Pittsburgh, PA. Los Alamitos, CA: IEEE Computer Society.

Holland, J. H. 1975. *Adaptation in Natural and Artificial Systems*. Ann Arbor, MI: University of Michigan Press.

Holland, J. H. 1986. Escaping brittleness: The Possibilities of General Purpose Learning Algorithms Applied to Parallel Rule-Based Systems. In *Machine Learning: An Artificial Intelligence Approach*, Volume 2, ed. R. S. Michalski, J. G. Carbonell, and T. M. Mitchell, 593–623. Los Altos, CA: Morgan Kaufmann.

Holsheimer, M., and A. P. Siebes. 1994. Data mining: The search for knowledge in databases. Technical report CS-R9406, Department of Computer Science, CWI, Amsterdam.

Hull, R., B. Kumar, A. Sahuguet, and M. Xiong. 2002. Have it your way: Personalization of network-hosted services. In *Proceedings of the 19th British National Conference on Databases*, ed. B. Eaglestone, S. North, and A. Poulovassilis, 1–10, Sheffield, UK. Berlin, Germany: Springer.

Hull, D., K. Wolstencroft, R. Stevens, C. Goble, M. Pocock, P. Li, and T. Oinn. 2006. Taverna: A tool for building and running workflows of services. *Nucleic Acids Research* 34:729–732.

Jeffery, K. 2003. GRIDs and ambient computing: The next generation. *Proceedings of the 5th International Workshop on Engineering Federated Information Systems*, ed. A. E. James, S. Conrad and W. Hasselbring, 2–13, Coventry, UK.

Johnston, W. E. 2002. Computational and data Grids in large-scale science and engineering. *Future Generation Computer Systems* 18(8):1085–1100.

Judd, D., K. McKinley, and A. K. Jain. 1996. Large-scale parallel data clustering. *Proceedings of the 13th International Conference on Pattern Recognition*, 488–493, Wien, Austria. Los Alamitos, CA: IEEE Computer Society.

Kargupta, H., R. Bhargava, K. Liu, M. Powers, P. Blair. S. Bushra, and J. Dull. 2003. VEDAS: A mobile and distributed data stream mining system for real-time vehicle monitoring. K. Sarkar, M. Klein, M. Vasa, and D. Handy. In *4th SIAM International Conference on Data Mining*, ed. M. W. Berry, U. Dayal, C. Kamath, and D. B. Skillicorn, Lake Buena Vista, FL. SIAM.

Kargupta, H., and P. Chan (eds.). 2000. *Advances in Distributed and Parallel Knowledge Discovery*. Palo Alto, CA: AAAI Press.

Kargupta, H., B. Park, D. Hershberger, and E. Johnson. 2000. A New Perspective Toward Distributed Data Mining. In *Advances in Distributed and Parallel Knowledge Discovery*, ed. H. Kargupta and P. Chan, 133–184. Menlo Park, CA: AAAI/MIT Press.

Kargupta, H., B. Park, S. Pitties, L. Liu, D. Kushraj, and K. Sarkar. 2002. MobiMine: Monitoring the stock marked from a PDA. *ACM SIGKDD Explorations* 3(2):37–46.

Keenoy, K., A. Poulovassilis, and V. Christophides (eds.). 2003. *Proceedings of the 1st IST Workshop on Metadata Management in Grid and P2P Systems: Models, Services and Architectures*, London, UK.

Knight, K. 1990. Connectionist ideas and algorithms. *Communications of the ACM* 33(11):59–74.

Kufrin, R. 1997. Generating C4.5 production rules in parallel. In *Proceedings of the 14th National Conference on Artificial Intelligence*, ed. B. Kuipers and B. L. Webber, 565–570, Providence, RI. Menlo Park, CA: AAAI/MIT Press.

Lackovic, M., D. Talia, and P. Trunfio. 2009a. A Framework for Composing Knowledge Discovery Workflows in Grids. In *Foundations of Computational Intelligence Vol. 6: Data Mining Theoretical Foundations and Applications, Studies in Computational Intelligence*, ed. A. Abraham, A. Hassanien, A. Carvalho, and V. Snášel, 345–369. New York: Springer.

Lackovic, M., D. Talia, and P. Trunfio. 2009b. A service-oriented framework for executing data mining workflows on Grids. *Proceedings of the 4th International Workshop on Workflow Management*, 72–79, Geneva, Switzerland. Los Alamitos, CA: IEEE Computer Society.

Li, X., and Z. Fang. 1989. Parallel clustering algorithms. *Parallel Computing* 11(3):275–290.

Mastroianni, C., D. Talia, and P. Trunfio. 2004. Metadata for managing Grid resources in data mining applications. *Journal of Grid Computing* 2(1):85–102.

McClelland, W. S., and D. E. Rumelhart. 1986. *Parallel Distributed Processing: Explorations in the Microstructure of Cognition* (Volumes 1 and 2). Cambridge, MA: MIT Press.

Moore, R. W. 2001. Knowledge-Based Grids: Two Use Cases. *GGF-3 Meeting*. Frascati, Italy.

Murata, T. 1989. Petri nets: Properties, analysis and applications, *Proceedings of the IEEE* 77(4):541–580.

Neri, F., and A. Giordana. 1995. A parallel genetic algorithm for concept learning. In *Proceedings of the 6th International Conference on Genetic Algorithms*, L. J. Eshelman, 436–443, Pittsburgh, PA. San Francisco: Morgan Kaufmann.

NGG3 Expert Group Report. 2005. Strategic future for European Grids: Next generation GRIDs based on SOKU—A new paradigm for service delivery and software infrastructure, Brussels, Belgium.

Olson, C. F. 1995. Parallel algorithms for hierarchical clustering. *Parallel Computing* 21(8):1313–1325.

Papazoglou, M. P., and D. Georgakopoulos. 2003. Service oriented computing. *Communications of the ACM* 46(10):25–28.

Papazoglou, M. P., P. Traverso, S. Dustdar, and F. Leymann. 2007. Service-oriented computing: State of the art and research challenges. *IEEE Computer* 40(11):38–45.

Pautasso, C., and G. Alonso. 2006. Parallel computing patterns for Grid workflows. *Proceedings of the Workshop on Workflows in Support of Large-Scale Science*, 1–10, Paris, France. Los Alamitos, CA: IEEE Computer Society.

Pearson, R. A. 1994. A Coarse-Grained Parallel Induction Heuristic. In *Parallel Processing for Artificial Intelligence 2*, ed. H. Kitano, V. Kumar, and C. B. Suttner, 207–226. Amsterdam: Elsevier Science.

Peltz, C. 2003. Web services orchestration and choreography. *IEEE Computer* 36(19):46–52.

Pittie, S., H. Kargupta, and B. Park. 2003. Dependency detection in MobiMine: A systems perspective. *Information Sciences* 155(3–4):227–243.

Prodromidis, A. L., P. K. Chan, and S. J. Stolfo. 2000. Meta-Learning in Distributed Data Mining Systems: Issues and Approaches. In *Advances in Distributed and Parallel Knowledge Discovery*, ed. H. Kargupta and P. Chan, 81–87. Menlo Park, CA: AAAI/MIT Press.

Quinlan, J. R. 1986. Induction of decision trees. *Machine Learning* 1(1):81–106.

Quinlan, J. R. 1987. Generating production rules from decision trees. In *Proceedings of the 10th International Joint Conference on Artificial Intelligence,* ed. J. P. McDermott, 304–307, Milan, Italy. San Francisco: Morgan Kaufmann.

Radcliffe, N. J., and P. D. Surry. 1994. Co-operation through hierarchical competition in genetic data mining. Technical report 94-09, Edinburgh Parallel Computing Centre, United Kingdom.

Ripeanu, M., A. Iamnitchi, and I. Foster. 2002. Mapping the Gnutella Network. *IEEE Internet Computing* 6(1):50–57.

Rivest, R. L. 1987. Learning decision lists. *Machine Learning* 2(3):229–246.

Schopf, J. M., M. D'Arcy, N. Miller, L. Pearlman, I. Foster, and C. Kesselman. 2005. Monitoring and discovery in a Web services framework: Functionality and performance of the Globus Toolkit's MDS4. Argonne National Laboratory Technical report ANL/MCS-P1248-0405, Lemont, IL.

Shafer, J., R. Agrawal, and M. Mehta. 1996. SPRINT: A scalable parallel classifier for data mining. In *Proceedings of the 22nd International Conference on Very Large Databases*, ed. T. M. Vijayaraman, A. P. Buchmann, C. Mohan, and N. L. Sarda, 544–555, Mumbai, India. San Francisco: Morgan Kaufmann.

Shaikh Ali, A., O. F. Rana, and I. J. Taylor. 2005. Web services composition for distributed data mining. *Proceedings of the Workshop on Web and Grid Services for Scientific Data Analysis*, 11–18, Oslo, Norway. Los Alamitos, CA: IEEE Computer Society.

Simpson, P. K. 1990. *Artificial Neural Systems: Foundations, Paradigms, Applications, and Implementations.* New York: Pergamon Press.

Skillicorn, D., and D. Talia. 2002. Mining large data sets on Grids: Issues and prospects. *Computing and Informatics* 21:347–362.

Smith, S. F. 1980. A learning system based on genetic adaptive algorithms. PhD thesis, University of Pittsburgh.

Smith, S. F. 1984. *Adaptive Learning Systems. Expert Systems, Principles and Case Studies.* London: Chapman & Hall.

Stankovski, V., M. T. Swain, V. Kravtsov, T. Niessen, D. Wegener, J. Kindermann, and W. Dubitzky. 2008. Grid-enabling data mining applications with DataMiningGrid: An architectural perspective. *Future Generation Computer Systems* 24(4):259–279.

Talia, D. 1993. Reti neurali: modelli ed algoritmi. *Rivista di Informatica* 22(2):127–144.

Talia, D. 2002. The Open Grid Services Architecture: Where the Grid meets the Web. *IEEE Internet Computing* 6(6):67–71.

Talia, D. 2012. Clouds meet agents: Toward intelligent Cloud services. *IEEE Internet Computing* 16(2):78–81.

Talia, D., and P. Trunfio. 2010a. How distributed data mining tasks can thrive as knowledge services. *Communications of the ACM* 53(7):132–137.

Talia, D., and P. Trunfio. 2010b. Mobile Data Mining on Small Devices through Web Services. In *Mobile Intelligence: Mobile Computing and Computational Intelligence*, ed. L. Yang, A. Waluyo, J. Ma, L. Tan, and B. Srinivasan, 264–276. New York: John Wiley & Sons.

Talia, D., P. Trunfio, and O. Verta. 2005. Weka4WS: A WSRF-enabled Weka toolkit for distributed data mining on Grids. In *Proceedings of the 9th European Conference on Principles and Practice of Knowledge Discovery in Databases*, ed. A. Jorge, L. Torgo, P. Brazdil, R. Camacho, and J. Gama. 309–320, Porto, Portugal. Berlin, Germany: Springer.

Talia, D., P. Trunfio, and O. Verta. 2008. The Weka4WS framework for distributed data mining in service-oriented Grids. *Concurrency and Computation: Practice and Experience*, 20(16):1933–1951.

Tan, P. N., M. Steinbach, and V. Kumar. 2006. *Introduction to Data Mining.* Reading, MA: Addison-Wesley.

Taylor, I., M. Shields, I. Wang, and O. Rana. 2004. Triana, applications within Grid computing and peer to peer environments. *Journal of Grid Computing* 1:199–217.

Tian, M., T. Voigt, T. Naumowicz, H. Ritter, and J. Schiller. 2004. Performance considerations for mobile Web services. *Computer Communications* 27(11):1097–1105.

van der Aalst, W. M. P. 1999. Formalization and verification of event-driven process chains. *Information and Software Technology* 41(10):639–650.

van der Aalst, W. M. P., A. H. M. ter Hofstede, B. Kiepuszewski, and A. P. Barros. 2003. Workflow patterns. *Distributed and Parallel Databases* 14(1):5–51.

Wang, F., N. Helian, Y. Guo, and H. Jin. 2003. A distributed and mobile data mining system. *Proceedings of the 4th International Conference on Parallel and Distributed Computing, Applications and Technologies*, 916–918, Chengdu, China. Los Alamitos, CA: IEEE Computer Society.

Wasserman, P. D. 1989. *Neural Computing: Theory and Practice*. New York: Van Nostrand Reinhold.

Widrow, B., and M. A. Lehr. 1990. 30 years of adaptive neural networks: Perceptron, madaline and backpropagation. *Proceedings of the IEEE* 78(9):1415–1442.

Willshaw, D. 1988. Neural systems and models. In *Proceedings of the European Seminar on Neural Computing*, ed. C. Zomzely-Neurath, 1–3, London, UK.

Witten, H., and E. Frank. 2000. *Data Mining: Practical Machine Learning Tools with Java Implementations*. New York: Morgan Kaufmann.

Workflow Management Coalition. 1999. Terminology and Glossary. Document Number WFMC-TC-1011—Issue 3.0.

Zahreddine, W., and Q. H. Mahmoud. 2005. An agent-based approach to composite mobile Web services. *Proceedings of the 19th International Conference on Advanced Information Networking and Applications*, 189–192, Taipei, Taiwan. Los Alamitos, CA: IEEE Computer Society.

Zaki, M. J. 1999. Parallel and distributed association mining: A survey. *IEEE Concurrency* 7(4):14–25.

Zaki, M. J. 2000. Scalable algorithms for association mining. *IEEE Transactions on Knowledge and Data Engineering* 12(3):372–390.

Index